NEUROSCIENCE RESEARCH PROGRESS

NEURONAL MIGRATION

DISORDERS, GENETIC FACTORS AND TREATMENT OPTIONS

NEUROSCIENCE RESEARCH PROGRESS

Additional books in this series can be found on Nova's website under the Series tab.

Additional e-books in this series can be found on Nova's website under the e-book tab.

NEUROSCIENCE RESEARCH PROGRESS

NEURONAL MIGRATION

DISORDERS, GENETIC FACTORS AND TREATMENT OPTIONS

ANDREW GIRARD
AND
LUCA MOREAU
EDITORS

Nova Biomedical Books
New York

For permission to use material from this book please contact us:
Telephone 631-231-7269; Fax 631-231-8175
Web Site: http://www.novapublishers.com

Additional color graphics may be available in the e-book version of this book.

LIBRARY OF CONGRESS CATALOGING-IN-PUBLICATION DATA

Neuronal migration : disorders, genetic factors, and treatment options / editors, Andrew Girard and Luca Moreau.
p. cm.
Includes bibliographical references and index.
ISBN 978-1-62100-059-4 (hardcover)
1. Neurons. 2. Cell migration. 3. Developmental neurobiology. I. Girard, Andrew. II. Moreau, Luca.
QP363.N4778 2011
612.8--dc23

2011030125

Published by Nova Science Publishers, Inc. † New York

Contents

Preface

Neuronal migration is the method by which neurons travel from their origin or birth place to their final position in the brain. In this book, the authors present topical research in the study of the disorders, genetic factors and treatment options of neuronal migration. Topics include neuronal migration disturbance in neurodevelopment disorders; candidate dyslexia susceptibility genes and disorders of neuronal migration; electrically guiding neuronal migration and membrane-cytoskeletal interactions in the regulation of neuronal migration in the adult brain.

Chapter I - Migration of neurons from the area of their generation to their final destination is a central event in brain development. During development, this process is highly sensitive to the effect produced by the exposure to different environmental factors like pathogenic infections, methylmercury, lead, alcohol, ionizing irradiation and drugs, among other. In fact, it has been established that disturbance of the normal process of the neuronal migration is the major cause of brain malformation, functional abnormalities and some behavioral deficit in both experimental animals and humans. Indeed, an increasing number of studies have suggested that disruption of neuronal migration is involved in the etiology of neurodevelopment disorders (NDDs), like schizophrenia, autism spectrum disorder, attention-deficit/hyperactivity disorder, and dyslexia, as well as cerebral palsy and mental retardation. Although it is not completely defined all causes of neuronal migration disruption and the relationship of these alteration with NDDs, it has been found that patients with these conditions present abnormalities in genes that encode proteins which regulate neuronal migration during development; it has been suggested that the structural, cytoarchitectural and functional brain abnormalities found in these patients might be due, at least partially, to an aberrant neuronal migration process. Among the agents that it is known produce neuronal movement alteration are included different environmental factors, nevertheless, it is not completely defined the role and particularly the molecular mechanisms by which these factors contribute in the etiology of NDDs. In this chapter the authors review the current knowledge regarding the relationship between neuronal migration and developmental disorders. In particular, they focus on the disruption of neuronal migration induced by different environmental factors (biological, chemical and physical) and the role of these effects as a risk factor for NDDs. Recent advances in the molecular mechanisms of disruption of neuronal migration by environmental factors and their relationship with NDDs will also be reviewed, and how this knowledge may contributing to a better understanding of the physiopathology of NDDs in which neuronal migration disruption is involved.

Chapter II - Subtle alterations in the development of the neocortex are associated with cognitive deficits in humans and other mammals. A large number of genes modulate the development of the neocortex, and research into the behavioral phenotype associated with specific gene manipulations is advancing rapidly. Recent findings include evidence that anomalies in several human genes, (*DYX1C1, DCDC2, KIAA0319, MRPL19, C2ORF3, ROBO1, KIAA0319L*), may be associated with an increased incidence of dyslexia. Concurrent research has shown that the rat homologs for a subset of these genes (*DYX1C1, DCDC2, KIAA0319, ROBO1*) modulate critical parameters of early cortical development, including neuronal migration. Recent studies have utilized animal models of genetic disruption, using in utero RNA interference to knockdown expression of *Kiaa0319* and *Dyx1c1* in developing rat pups. Analyses have shown that interference with these genes leads to aberrant neuronal migration in the neocortex, resulting in the formation of ectopias and heterotopias in white matter. Interestingly, these malformations are similar to those observed in histological studies of brain tissue obtained *post mortem* from dyslexic individuals. Additionally, these genetic disruptions to neuronal migration have been associated with deficits in rapid auditory processing, spatial learning, and working memory in animal models. Parallel deficits in core processes critical to language and learning have been reported in dyslexic populations. Specifically, whereas dyslexia is diagnostically defined overall as a reading impairment in the absence of any generalized cognitive or neurologic deficit, this disorder likely reflects complex effects of multiple genes on multiple neural systems. As such, it is not surprising that more fine-grained behavioral sub-phenotypes have been identified within the dyslexic population (also called "intermediate phenotypes" in human genetic studies; see Marino *et al.*, 2007), and these may ultimately prove easier to link to the effects of specific dyslexia-risk genes. Reported sub-phenotypes include specific deficits in short term or working memory (verbal and also non-verbal), visual attention, and rapid auditory processing. It must be noted that the mapping of these intermediate phenotypes onto "clinical sub-types" of dyslexia (e.g., phonologic versus surface or orthographic dyslexia) has proved difficult—likely due to variations in definition, classification, and approach—but it seems likely that the clinical sub-types reported for dyslexia may reflect differential intermediate phenotypes and, in turn, the influence of different genes or gene combinations. One benefit of a focus on these intermediate phenotypes is that such behavioral measures are more amenable to modeling in animals (whereas reading *per se* is not). Thus, rodent models can be used in paradigms of working memory, visual attention and rapid auditory processing. Moreover, such rodent models can be employed to link specific intermediate phenotypes with individual dyslexia-risk genes. In the present review, behavioral results and histological data are discussed, and comparisons are made between the neuroanatomical and behavioral impacts of early interference with *Dyx1c1* and *Kiaa0319* in rodents. The recent findings are summarized and discussed in the context of improving the authors' understanding of the genetic contributions to the behavioral expression of dyslexia in humans.

Chapter III - Guided and stimulated nerve growth and neuronal migration is critical to repairing diseased or damaged nervous systems. Electric fields (EFs) provide a powerful signal to stimulate and guide nerve growth and neuronal migration. EFs may direct axon outgrowth, orientate cell polarity, and guide neuronal migration. These effects may be mediated through activation of important intracellular signaling pathways and orientation of important organelles. The potential use of EFs in the nervous system may provide new therapies for neurodevelopmental disorders and brain injuries. This chapter aims to review recent experimental data on EFs-guided nerve growth and neuronal migration, and possible molecular and cellular mechanisms.

Chapter IV - At least two areas of the adult mammalian brain continue to produce new neurons throughout the adult lifespan: the hippocampus and the olfactory bulb. These new neurons are derived from stem cells located in the subgranular zone of the hippocampus, and the anterior portion subventricular zone surrounding the lateral ventricles, respectively. In the hippocampus, precursor cells divide in the subgranular zone and migrate relatively short distances into the overlying granule cell layer, where they differentiate into new granule cell neurons. In contrast, neuronal precursors destined for the olfactory bulb must migrate long distances first tangentially through the rostral migratory stream, and then radially into the outer layers of the olfactory bulb where they differentiate into two types of inhibitory interneurons: granule cells, and periglomerular cells. This migration requires an ongoing rearrangement of the cytoskeleton as well as integration of chemo-attractive and repulsive cues from the environment to result in directed movement of the cell along the appropriate pathways. Molecules proposed to play a role in the migration of neuronal precursors along the rostral migratory stream include polysialated neural cell adhesion molecule, netrins, semaphorins, slits, ephrins, and neuregulins. In addition, two molecules in the extracellular matrix, reelin and Tenascin-R, as well as the secreted factor prokineticin-2 have been proposed to key roles in the transition from radial to tangential migration. In this review, the author discusses what is known about the factors regulating the migration of adult neuronal precursors to the olfactory bulb, as well as recent unpublished data from the author's laboratory suggesting possible roles of members of the ERM (ezrin-radixin-moesin), cadherin, and catenin families in this process.

Chapter V - Ischemic stroke is a leading cause of death and long-term disability worldwide. However, repairing damaged brain tissue is not feasible with current treatments. In recent years, it became clear that specific regions of the adult brain such as the subventricular zone (SVZ) and the subgranular zone (SGZ) are capable of responding to various stimuli including injurious signals from other brain areas by increasing the proliferation of resident neural progenitors. Progenitor proliferation and the migration of the newly formed neural progenitors towards the injured sites are regarded as an important endogenous repair process although current evidence indicates that the proliferation and migration activities are far from sufficient for satisfactory tissue repair. Thus, understanding the endogenous regenerative mechanism including proliferation and migration of neural progenitors is imperative for promoting tissue repair and regenerating an ischemia-injured brain structure.

Chapter VI - Correct neuronal migration is a key event for normal development and functioning of the central nervous system (CNS). In the mammalian cerebral cortex neurones ultimately arrange into a complex laminar configuration consisting of six distinguishable layers. Much has been discovered concerning the molecular and cellular mechanisms that underlie the correct movement of immature neurones within the cerebral cortex, as well as the crucial role played by the radial glial scaffold. In recent years, the importance of the interaction between the radial glial end-feet and the pial basement membrane has become increasingly apparent, and it is this aspect which forms the focus of the present review. The authors begin by outlining some of the key processes involved in the organisation of the cerebral cortex, focusing on several recent breakthroughs particularly the role of cytoskeletal and signalling proteins. They discuss basement membranes, specifically the role played by the pial basement membrane in maintaining the correct laminar structure of the developing cortex. Finally the authors end with a discussion on how animal models have contributed to their understanding of cortical malformations, such as cobblestone lissencephaly and speculate on developments that may ultimately offer avenues for future therapeutic interventions.

In: Neuronal Migration
Editors: Andrew Girard and Luca Moreau
ISBN: 978-1-62100-059-4

Chapter I

Neuronal Migration Disturbance in Neurodevelopment Disorders: Role of Environmental Factors

Daniela Arias-Salvatierra[1], Leonor C. Acosta-Saavedra[1], Damaris Albores-García[1], Ellen Silbergeld[2], and Emma S. Calderon-Aranda[1*]

Departamento de Toxicología, Centro de Investigación y de Estudios Avanzados DF, México

Department of Environmental Health Sciences, Bloomberg School of Public Health, Johns Hopkins University, Baltimore, Maryland, US

Abstract

Migration of neurons from the area of their generation to their final destination is a central event in brain development. During development, this process is highly sensitive to the effect produced by the exposure to different environmental factors like pathogenic infections, methylmercury, lead, alcohol, ionizing irradiation and drugs, among other. In fact, it has been established that disturbance of the normal process of the neuronal migration is the major cause of brain malformation, functional abnormalities and some behavioral deficit in both experimental animals and humans. Indeed, an increasing number of studies have suggested that disruption of neuronal migration is involved in the etiology of neurodevelopment disorders (NDDs), like schizophrenia, autism spectrum disorder, attention-deficit/hyperactivity disorder, and dyslexia, as well as cerebral palsy and mental retardation. Although it is not completely defined all causes of neuronal migration disruption and the relationship of these alteration with NDDs, it has been found

* Corresponding author: Dr. Emma S. Calderon-Aranda, Toxicología, Cinvestav, IPN, P.O. Box 14-740, México, D.F. 07360, México. Phone: (52) (55) 57473800 EXT. 5473; Fax: (52) (55) 57473395; e-mail: scalder@cinvestav.mx.

that patients with these conditions present abnormalities in genes that encode proteins which regulate neuronal migration during development; it has been suggested that the structural, cytoarchitectural and functional brain abnormalities found in these patients might be due, at least partially, to an aberrant neuronal migration process. Among the agents that it is known produce neuronal movement alteration are included different environmental factors, nevertheless, it is not completely defined the role and particularly the molecular mechanisms by which these factors contribute in the etiology of NDDs. In this chapter we review the current knowledge regarding the relationship between neuronal migration and developmental disorders. In particular, we focus on the disruption of neuronal migration induced by different environmental factors (biological, chemical and physical) and the role of these effects as a risk factor for NDDs. Recent advances in the molecular mechanisms of disruption of neuronal migration by environmental factors and their relationship with NDDs will also be reviewed, and how this knowledge may contributing to a better understanding of the physiopathology of NDDs in which neuronal migration disruption is involved.

Introduction

Brain development represents a critical period of vulnerability for the establishment of adult mental health. From the prenatal period through the first years of life and until adolescence, the brain undergoes its most rapid development, and is also most susceptible to be influence by external experiences. There is now compelling evidence that environmental factors during prenatal or early neonatal life alter development and predispose the individual to lifelong health problems, including susceptibility to NDDs (Cirulli et al., 2009; Laviola et al., 2009). A NDD is defined as an impairment of the growth and development of central nervous system (CNS). It affects an individual's brain function in controlling emotion, learning ability, memory and social interaction. The term NDDs is used in two ways: the first refers to conditions affecting children's neurological development with a known genetic or acquired etiology, such as fragile X syndrome or fetal alcohol spectrum disorders (FASDs); the second refers to conditions of presumed multi-factorial etiology in which certain aspects of neurodevelopment are selectively impaired; this includes such conditions as schizophrenia, autism spectrum disorder (ASD), attention deficit hyperactivity disorder (ADHD), and developmental dyslexia (Bishop and Rutter, 2008).

NDDs are suggested to affect millions worldwide, with significant financial and emotional burdens for patients, their families, and society (NIH, 2000). The epidemiology of NDDs differs in a case by case basis and varies by country. For example, the prevalence estimated for ASD is 0.6% (Chakrabarti and Fombonne, 2005) and the world estimated prevalence for ADHD is 5.29% (Polanczyk et al., 2007). Although, this prevalence represent the average for industrialized countries even though almost 90% of the world's children are born in developing countries.

Even though the cause for the vast majority of NDDs remains unknown, it is generally accepted that the cause is likely to include the alteration of genes that regulate developmental processes such as cell proliferation, differentiation, apoptosis, migration and environmental insults that aggravate these effects on neurodevelopment processes. However, it is not clear yet whether environmental factors act merely as precipitating agents, always requiring genetic liability, or whether they represent an essential component of the pathogenesis process in which genetic liability alone does not lead to the full range of NDDs. In this chapter we will discuss the role of neuronal migration in NDDs and the susceptibility of this process to environmental insults and to point out the participation of this susceptibility in the associated risk of NDDs.

On the other hand, the developmental exposure to environmental factors could interfere with proliferation, migration, differentiation, synaptogenesis, among others, being the specific process altered related to timing and dose of exposure. Even tough in the etiology of NDDs it has been involved the interaction between genetic influences and environmental factors that affect brain maturational processes (Lewis, 2001), and particularly disrupt the neuronal migration process during susceptible windows of the neurodevelopment, the molecular mechanisms by which each pollutant exerts these effects are not completely understood. In this chapter it is presented a review of recent evidence on infection, methylmercury (MeHg), lead (Pb) and alcohol, some of the more known environmental pollutants that affect neuronal migration. We focus in the evidence of alterations in neuronal migration, as well as insights in the molecular mechanisms involved in these alterations. Indeed, it is important to note that neuronal migration process involves complex interactions between glial and neuronal cells (Kriegstein 2005; Stipursky and Gomes 2007), different growth factors, cytoskeleton, cell adhesion molecules, extracellular proteins, as well as multiple signaling pathways including growth factors, cytokines and chemokines, cytokine's and chemokine's receptors, cell adhesion molecules among other proteins (Lee and Benveniste, 1999; Szelényi, J., 2001; Clegg et al., 2003; Cartier et al., 2005; Stumm et al., 2007). Because the large number of proteins involved in regulation of neuronal migration in physiological conditions, the identification of molecular targets and molecular mechanisms of environmental factors in neuronal migration disruption is not completely understand. In this chapter are present recent advances in the molecular mechanisms of disruption of neuronal migration by environmental factors, and how the knowledge of molecular mechanisms may contribut to a better understanding of the physiopathology and early detection of NDDs in which neuronal migration disruption is involved.

1. Neuronal Migration in Neurodevelopment Disorders

Migration of neurons from the proliferative zones to their final destination in the mature nervous system is an essential event for the development of nervous system (Valiente and Marín, 2010). Once an immature neuron is generated, it migrates from the ventricular zone towards the pial surface along radial glial processes and then settles in specific cortical layers (Rakic et al., 2007). This process requires molecules from the cytoskeleton which play important role in the initiation and ongoing progression of neuronal movement; signaling molecules playing a role in lamination, molecules modulating glycosylation which seem to provide stop signals for migrating neurons, neurotransmitters and trophic factors (Marin and Rubenstein, 2003). Regulation of the timing, onset, direction and stop of neuronal migration is highly ordered. Disturbances of the normal process of neuronal migration are clearly responsible for the neuronal migration disorders which include lissencephaly, heterotopias, polymicrogyria, schizencephaly and focal cortical dysgenesis (FCD). FCD is characterized by malformation owing to anomalies of cortical development leading to severe mental retardation and epilepsy (Spalice et al., 2009). Several genetic defects cause neuronal migration disorders, including mutation in *Lis1* and *doublecortin* genes (Mochida and Walsh, 2004). However, it is possible that subtle alterations in the molecules associated with neuronal migration could be associated with other NDDs. The disruption of neuronal migration has been suggested is involved in the etiology of NDDs, especially in schizophrenia and ASD.

Schizophrenia

In recent years the schizophrenia has been proposed is an outcome of an aberration in neurodevelopment processes that begins long before the onset of clinical symptoms and is caused by a combination of environmental and genetic factors (Rapoport et al., 2005). Early postmortem studies have shown neuronal disarray, especially in lamina II of the entorhinal cortex, and abnormal migration of subplate neurons in the neocortical white matter, which were strongly supportive of prenatal neurodevelopment abnormalities particularly in neuronal migration and organization (Jakob and Beckmann, 1986; Akbarian et al., 1993; Arnold et al., 1991; Arnold, 1999). Recently, it was found that schizophrenic patients present increased levels of chondroitin sulfate proteoglycans-positive-glia in the deep amygdale nuclei (419-1162%) and entorhinal cortex (480-1560%) (Pantazopoulos et al., 2010). Chondroitin sulfate proteoglycans are one of the main components of the extracellular matrix that function as organizers of the neural extracellular matrix in the adult and have an important role in the control neuronal migration during development (Maeda et al., 2010). The disruption of these interactions may represent a factor contributing to disturbances of the neuronal migration with synaptic connectivity and GABAergic, glutamatergic and dopaminergic neurotransmission observed in schizophrenia. The application of microarray technology in studies with postmortem brains has also added support to the hypothesis of the participation of neuronal migration disturbances in schizophrenia. For example, one of the most consistent findings is the reduced level of the 67 kDa glutamic acid decarboxylase (GAD67) and several studies reported positive associations between its gene *GAD1* and schizophrenia (Akbarian and Huang, 2006). Preclinical studies show that GAD67 is involved with cortical, thalamic, cerebellar and hippocampal development (Maqueda et al., 2003, Frahm and Draguhn, 2001). In fact, GAD67 mRNA expression shows a maximum of expression during the period of neuronal migration and differentiation in the brain (Szabo et al., 1994) and attenuated neuronal migration of GnRH-1 neurons to the nasal/forebrain junction were observed in GAD67 knockout mice during embryonic development (Lee et al., 2008). The effect on GAD67 and the decreasing of reelin levels, a secreted extracellular matrix protein required for proper neuronal migration (Zhao and Frotscher, 2010), are two of the more consistent findings described in postmortem material from schizophrenic patients (Torrey et al., 2005). In fact, reduced reelin in brains and alterations of reelin levels in blood has been found in schizophrenic patients in contrast to control subjects (Fatami et al., 2000; Fatemi et al., 2001).

Neuroregulin (*NRG1*) is another susceptibility gene for schizophrenia involved in neuronal migration. Polymorphisms in the *NRG1* and its receptor *ErbB4* are associated with schizophrenia and loss of function of NRG1, ErbB4; perturbation of the NRG1 signaling can cause deficits in the migration of pyramidal and GABAergic neurons, neurite outgrowth and axon projection (Mei and Xiong, 2008), which is consistent with the abnormal neural developmental model of schizophrenia.

Autism Spectrum Disorder (ASD)

ASD are complex neuropsychiatric disorders of neurodevelopmental origin that includes autistic disorder, Asperger syndrome and pervasive developmental disorder not otherwise specified. The phenotypic heterogeneity of ASD suggest that variations are caused by different genes associated with alterations in different developmental processes in specific brain regions and are related to a diverse cognitive impairments and functional abnormalities (Happé et al., 2006). Disease mechanisms may include altered increased cell proliferation, decreased cell death, altered synapse elimination, and altered cell migration (Keller and Persico, 2003). Postmortem analysis found subcortical, periventricular, hippocampal and cerebellar heterotopias in the brains of autistic subjects reflecting abnormal neuronal migration (Wegiel et al., 2010).

As in schizophrenia, genes associated with ASD are known to regulate neuronal migration during development. Met is a tyrosine kinase receptor activated by hepatocyte growth factor (HGF). During development Met-signaling regulates a range of cellular processes including neuronal migration (Sun et al., 2002; Segarra et al., 2006). Altered regulation of human *MET* (gene coding for Met) expression has been implicated in autism. A single nucleotide polymorphism (SNP) in the *MET* promoter that disrupts the transcription of *MET* has been reported to be strongly associated with ASD in Caucasian (Campbell et al., 2008) and a SNP located in the intron 1 of *MET* was also associated with ASD in Japanese subjects (OR =1.61 P=0.044) (Thanseem et al., 2010). Furthermore, transmission of the *MET* allelic variant from parent to child is associated with social and communication phenotypes observed in ASD (Campbell et al., 2010). Analyses of postmortem cerebral cortex from ASD patients revealed altered expression of *MET* and transcripts encoding proteins that regulate Met signaling, HGF, urokinase plasminogen activator receptor (PLAUR) (Campbell et al., 2007); indeed, a promoter variant in *PLAUR* gene, has also been associated with ASDs by case-control and family-based association studies (Campbell et al., 2008).

Reelin alterations are also observed in ASD; reduced levels of reelin has been observed in postmortem brain cortices of autistic patients (Fatemi et al., 2005) and in blood of autistic twins, their fathers, their mothers and their phenotypically normal siblings versus controls (Fatemi et al., 2002). Analysis of SNPs in the *reelin* gene found significant differences in the transmission of the alleles of exon 22 and intron 59 SNP in autistic subjects suggesting a possible association between SNPs in reelin and autism (Serajee et al., 2006).

Attention-deficit/hyperactivity Disorder (ADHD)

ADHD is a heterogeneous developmental disorder characterized by early onset of age-inappropriate persistent and pervasive symptoms of attention, hyperactivity and impulsivity (Faraone et al., 2005). Brain magnetic resonance imaging (MRI) scans of children with ADHD revealed an augment in the frequency of gray matter heterotopias and neuronal migration anomalies (Nopoulos et al., 2000). Recently, genome-wide association studies

(GWAS) for ADHD revealed a consistent association between the gene *CDH13* and *ADHD* (Franke et al., 2009; Banaschewski et al., 2010); *CDH13* is located on chromosome 16q24 within the only significant linkage peak from a meta-analysis of seven linkage studies in *ADHD* (Zhou et al., 2008). *CDH13* encodes cadherin 13, a protein expressed in neurons that is involved in cell migration and cell-cell communication (Andreeva and Kutuzov, 2010).

Developmental Dyslexia

Dyslexia is a learning disability in which reading (but not any other impairment) is the most prominent symptom. The brains of individuals with developmental dyslexia have neocortical neuronal migration abnormalities including molecular layer heterotopias, laminar dysplasia and periventricular nodular heterotopias (Chang et al., 2005). Using MRI it was found that patients with periventricular nodular heterotopias have heterotopic nodules containing misplaced neurons that would normally have migrated to the cortex, suggesting the participation of disruption of neuronal migration on dyslexia (Walker et al., 2009). Association studies found that the *DYX1* locus corresponding to 15q21, correlates with reading ability and spelling, this finding served to identify the *DYX1* locus as a candidate for dyslexia susceptibility (Grigorenko et al., 1997; Morris et al., 2000). The *DYX1C1* gene is near the *DYX1* locus in chromosome 15q21, disrupted by translocation and segregated with dyslexia (Taipale et al., 2003). Two sequence changes in *DYX1C1* are associated with dyslexia and with relevant functional consequences like decreasing gene expression and producing a truncated unfunctional protein (Taipale et al., 2003; Tapia-Páez et al., 2008). In the adult human brain, DYX1C1 protein localizes in a fraction of cortical neurons and white matter glial cells (Taipale et al., 2003). Knockdown of the homologue gene *Dyx1c1* in rats during embryonic life dramatically disrupts neuronal migration of the cerebral cortex with hippocampal pyramidal cell migration anomalies and molecular layer ectopias (Rosen et al., 2007).

The structural, cytoarchitectural and functional brain abnormalities reported in patients with schizophrenia, ASD, ADHD, and dyslexia might be due in part to aberrant neuronal migration, since the final position of neurons affects neuronal function, morphology, and formation of synaptic connections (Chevassus and Represa, 1999; Greessens, 2005). However, it is not clear if neuronal migration disruption is a factor strong enough to cause the behavioral effects observed in these NDDs.

2. Environmental Factors that Alter Neuronal Migration: Role in Neurodevelopment Disorders

Genetic and postmortem brain analyses suggest a relevant role for neuronal migration disruption in the etiology of NDDs. Recent research is starting to elucidate the effects of environmental insults on neurodevelopmental processes in particular on neuronal migration, as well as the relevance of these effects in the etiology of NDDs. Some of the most studied environmental factors potentially associated to NDDs are briefly revised below.

Infections

Substantial evidence over the last decade suggests that activation of the immune system during pregnancy increases the risk of the offspring for NDDs. The observation that influenza epidemics may be followed by a transient increase in births of schizophrenic patients led to the hypothesis that maternal viral infections contribute to the etiology of schizophrenia (Sham et al., 1993). Epidemiological studies found that exposure to influenza four months prior to birth (i.e., about the 6th month of gestation) was significantly associated with an increased risk of later schizophrenia; however, according to the calculated population, 1.4% of the whole schizophrenic sample population was attributed to prenatal exposure to influenza (Takei et al., 1996). Later, other two viral infections *in utero*, rubella and respiratory infection were associated with later development of schizophrenia (Brown and Susser, 2002). This association has also been found with *Toxoplasma gondii* infection during pregnancy determined through sera antibody in mothers of schizophrenic patients (OR 2.61) (Mortensen et al., 2007).

The relation between infections and schizophrenia has also been supported through the associative studies between deficits and morphological characteristic observed in this NDD. *In utero* infection was associated with increased length of the cavum septum pellucidum (a neurodevelopmental morphologic anomaly often found in schizophrenia) in schizophrenic patients (Brown et al., 2009). Additionally, cognitive deficits observed in schizophrenia are also associated with prenatal exposure to maternal infection; executive dysfunction is one of the most prominent and functionally important cognitive deficits found in schizophrenia. Regarding that, schizophrenic patients who were exposed to infection *in utero* (influenza and toxoplasmosis) had a poorer performance in executive function tests in comparison to unexposed patients (Brown et al., 2009). Another interesting finding that supports the role of maternal immune activation in the etiology of schizophrenia, is the fact that has been found high levels of interleukin 8 (IL-8) in the serum of mothers with schizophrenic offspring obtained during the second-trimester of gestation in comparison to control subjects (Brown et al., 2004). IL-8 is a cytokine belonging to a family of soluble polypeptides critical in the immune response to infection and other inflammatory process. Recently, it was found in schizophrenic patients that fetal exposure to increased levels of IL-8 (evaluated in archived maternal prenatal sera) were associated with an increase in ventricular cerebrospinal fluid and with decreases in left entorhinal cortex volumes, while no significant associations were observed among controls. This is an exciting finding, since previous work found neuronal disarray in the entorhinal cortex associated with abnormal neuronal migration in schizophrenic patients (Jakob and Beckmann, 1986; Akbarian et al., 1993; Arnold et al., 1991; Arnold, 1999); these results support the relevant role of cytokines in neuronal migration during development (Garay and McAllister et al., 2010).

The possible association of ASD with infection was proposed in early studies finding an increased frequency in births of autistic children during the months with more frequent bronchitis/pneumonia hospitalization (Tanoue et al., 1988). Recently, it was observed that this association is dependent of timing; admission to hospital due to maternal viral infection in the first trimester and maternal bacterial infection in the second trimester; actually, in this study were found association of infection with diagnosis of ASDs in the offspring [adjusted hazard ratio = 2.98 (CI: 1.29-7.15) and adjusted hazard ratio = 1.42 (CI: 1.08-1.87), respectively]

(Atladóttir HO et al., 2010). Altered inflammatory molecules are also observed in individuals with ASD; elevated levels of pro-inflammatory cytokines IL-6, TNF-α and IL-1β have been found in the cerebrospinal fluid, blood, and post-mortem brain tissue of individuals with autism (Vargas et al., 2005). Moreover, microglial and astrocyte activation is concurrent with cytokine up-regulation in the autistic brain and cerebrospinal fluid (Vargas et al., 2005).

Increased risk for ADHD was also observed in association with maternal infections during pregnancy; maternal genitourinary infection was associated with significantly increase risk of ADHD (OR = 1.29, 95% CI = 1.23-1.35); the diagnosis of specific pathogen showed that infection by *Chlamydia/nongonococcal*, trichomoniasis, and candidiasis were associated with increased risk for ADHD, except gonorrhea (Mann and McDermott, 2010). Additionally, it has been determined in a case-control study that an increased risk of ADHD is present in children born to women experiencing a viral rash (i.e. measles, varicella, or rubella) during pregnancy (Arpino et al., 2005).

Infections during development could increase the risk for NDDs through altering developmental process such as neuronal migration. Infection of pregnant mice with influenza on embryonic day 9.5 cause heterotopic Purkinje cells as well as delayed migration of granule cells in cerebellar lobules VI and VII of their offspring. This observed pathology in the offspring of influenza exposed mice is similar to that observed in autism (Shi et al., 2009).

On the other hand, cytomegalovirus (CMV) is the most significant infectious cause of congenital anomalies in the CNS associated to intrauterine infection in humans with an incidence of approximately 5-10% of live births (Tsutsui et al., 2009). In children congenital CMV infection has been associated with the development of typical autistic disorder (Yamashita et al., 2003). CMV infection affects the developing CNS through disruption of migration of neural cells, resulting in brain disorders. Injection of mouse CMV into the cerebral ventricles of mouse embryos caused a disturbance of neuronal migration and a marked loss of neurons in postnatal brains (Shinmura et al., 1997).

Methylmercury

An organic form of mercury (Hg), MeHg, is the most common form of Hg-exposure to the population, mainly through the consumption of fish and the most toxic for the developing SNC (Holmes et al., 2009). The effect of MeHg on developing CNS is dependent on the dose, duration and the developmental stage of exposure. Prenatal exposure may result in irreversible damage to the fetal CNS. *In utero* exposure to high doses such those seen in episodes of Iraq and Japan (Hg maternal-pregnancy levels in blood ranging from 40-50 μg/l) provoked cerebral palsy, blindness, deafness, cerebellar ataxia, impaired physical growth and severe mental retardation (NRC, 2000). At low levels of exposure, MeHg is associated with deficits on attention, memory and language (Stein et al., 2002). Prenatal exposure to MeHg through maternal fish consumption was investigated in three cohort studies conducted in the Faroe Islands, New Zealand and the Republic of Seychelles. The average concentrations in maternal hair in these studies were 4.3 μg/g, 8.3 μg/g, and 5.8 μg/g, respectively (Cohen et al., 2005). The Faroe Islands study reported a negative dose-response relationship between the results obtained by children on standardized neurobehavioral tests and consumption of seafood contaminated with MeHg by their mothers during pregnancy. Deficits were observed in tests of attention, fine motor function, visual and spatial ability and verbal memory. The New Zealander study in pregnant women found similar associations between fish consumption and neurodevelopment effects on children. However, the Seychelles study did not (WHO, 2004). At geometric mean blood Hg concentration of 2.4 μg/l were not found significantly associated with ADHD symptoms in children from the Children's Health and Environment Research survey in Korea (Ha et al., 2009). In spite of the effects on CNS development in children above described, the association between MeHg and NDDs is inconclusive, and more studies are required.

Ethylmercury in the form of thimerosal is an organic form of Hg used in the vaccines as an antiseptic and antifungal agent as a preservative. In a retrospective cohort study in the United Kingdom was evaluated if exposure to thimerosal (defined according to the number of vaccine doses containing thimerosal received in the first semester of life) was associated with ADHD, autism, language or speech delay, tics and behavior problems; only for tics was evidence of a higher risk (hazard ratio: 1.5 per dose, 95% CI 1.02-2.20), whereas a statistically significant negative association was found with increasing doses for ADHD (hazard ratio: 0.79; 95% CI: 0.69-0.98) (Andrews et al., 2004). However, a study using the same measure of exposure (doses received of Hg from vaccines containing thimerosal in the first semester of life) found significantly increased risk for autism, speech disorder, mental retardation and personality disorders in a two phase population-based epidemiological study in the United States (Geier and Geier, 2005). On the other hand, graphic ecologic analyses using population-based data from the United States, Sweden and Denmark found no association between the prevalence/incidence of autism with average exposures to thimerosal-containing vaccines (Stehr-Green et al., 2003).

The fact that Hg received so much attention in concern of NDDs is that Hg can alter developmental processes implicated in the pathology of NDDs. A regular finding in autism is an augment in the number of reactive glia in the CNS, as occur in subjects exposed *in utero* to MeHg; the cerebellum is implicated in the pathology of autism with effects on Purkinje and granular cells, and in subjects exposed to MeHg the cerebellum is an important target; it has been found that *in utero* exposure to MeHg alters neuronal migration, disruption that has been involved in NDDs. Another interesting finding is the ability of Hg and MeHg to cause autoimmune responses and the fact that an autoimmune component has been implicated in the etiology and pathology of autism (Cohly and Panja, 2005).

Morover, it was found that autistic children had a higher Hg exposure during pregnancy due to maternal dental amalgam and thimerosal-containing vaccine injections (Mutter et al., 2005); subjects with more or equal to 6 maternal amalgams were diagnosed with autism (severe) and were found to increase the risk (OR: 3.2) of autism (severe) in comparison to ASD (mild) in subjects with less OR equal to 5 maternal amalgams during pregnancy (Geier et al., 2009). Then, it is possible that Hg exposure by itself is not able to conduct to a NDD. Instead, the relationship between Hg exposure and NDDs probably could be part of several developmental insults interacting (infection and Hg, or Hg and immune genetic susceptibility) for the development of NDDs.

Lead

Recently, association between Pb exposure and ADHD has also been found. Lead in blood (PbB) has been associated with higher scores in the ADHD index (beta = 0.17; p= 0.05) in children 3-7 years of age attending pre- and elementary schools in India (Roy et al., 2009). In a large population-based study of 4700 American children aged 4-15 years was found a significant dose-dependent association between the level of PbB and a clinical diagnosis of ADHD (Braun et al., 2006). The children had a 4 fold increased risk of ADHD when PbB levels were >2.1 μg/dL. The association between PbB levels and ADHD was also found in Korean and Chinese children (Ha et al., 2009; Wang et al., 2008). Of concern is the finding that a variety of insults can participate to increase the risk for ADHD. Froehlich et al. (2009) found that children exposed to Pb (current PbB concentrations) and prenatal tobacco exposure (report of maternal cigarette use during pregnancy) had a greater risk of ADHD [OR 8.1 (95% CI: 3.5-18.7)], in comparison to tobacco exposure [2.4 (95% CI: 1.5-3.7)] and PbB concentrations alone [OR for third versus first tertile: 2.3 (95% CI: 1.5-3.8)]. Using δ-aminolevulinic acid (from maternal serum samples taken in the second trimester of gestation) as a biologic marker of Pb exposure in a case control study it was found that schizophrenic patients had a OR=2.43 of having higher δ–aminolevulinic acid, which correspond to PbB levels higher than or equal to 15 μg/dl (Opler et al., 2004).

Alcohol

Partial fetal alcohol syndrome (FAS), alcohol related birth defects and alcohol-related NDDs are part of FASD associated to maternal alcohol consumption during pregnancy. In FASD the most profound effects of prenatal alcohol exposure are on the developing brain and the cognitive and behavioral effects that ensue. The effects include cognitive and behavioral disabilities irreversible as a result of abnormal brain development, pre- and postnatal growth retardation and facial dysmorphism (Ramsay, 2010). Alcohol exposure affects brain development through the disruption of processes from neurogenesis to myelinization depending on developmental stage, dose and frequency of exposure (Ramsay, 2010). In particular, *in utero* exposure to alcohol can result in disrupted neuronal migration with structural abnormalities of the brain and ectopic neurons often found in the brains of FAS patients (McGee and Riley, 2006). In 1978, Clarren et al. (1978) described the neuro-pathological findings in four human neonates who were exposed to large quantities of alcohol at frequent intervals during gestation; they found that brains displayed similar malformations stemming from error in migration of neuronal and glial elements. Wisnieswski et al. (1983) found microencephaly and parenchymal heterotopias at postmortem analysis of five patients diagnosed with FAS (ages from 2 days to 4months). Analysis by structural MRI revealed brain hypoplasia and brain microcephaly in patients with FAS (Archibald et al., 2001). Another point to visualize the participation of neuronal migration disruption in FASD is that epilepsy/seizures have remarkably high prevalence in the FASD population (Guerrini and Barba, 2010). In a retrospective study conducted in all active charts at two FASD clinics, it was found that epilepsy and seizures had a prevalence of 17.7% (Bell et al., 2010).

Alcohol exposure during development has been associated with NDDs. The attention and inhibition problems found in children with ADHD are also common in children with FASD. ADHD is diagnosed in up to 94% of individuals with heavy prenatal alcohol exposure while the prevalence of ADHD in children in the general population ranges from 5% to 11% (O'Malley and Nanson, 2002) and alcohol exposure during development is an environmental risk factor for ADHD (RACP, 2009). In a cohort of 71 adopted children from Eastern Europe with history of prenatal alcohol exposure, it was found that 52% of the children were identified with FASD, 9% of this children were diagnosed for ASD, and 51% for ADHD. The children diagnosed as having FASDoften (95%) had comorbid neurodevelopment cognitive diagnoses, such as cognitive dysfunction, ADHD, mental retardation and autism (Landgren et al., 2010).

3. Molecular Mechanisms of Disruption of Neuronal Migration by Environmental Factors

Infections

In the developing CNS, treatment with lipopolysaccharide (LPS) has been used to evaluate the effects of inflammation on brain function because there is evidence that inflame-matory stimulus during CNS development notably increases the risk for adverse effects in the neonate and adult brain (Goncalves et al., 2002). However, the mechanisms responsible of these alterations are not well defined. LPS is a component of the outer membrane of Gram negative bacteria that binds to the host Toll-Like Receptor 4 (TLR-4) triggering inflammatory pathways. In fact, a variety of cognitive, pharmacological and behavioral alterations have been found in adult rodents following prenatal or neonatal exposure to LPS and in these models, neuronal precursors migrate to areas of LPS activation (Romero et al., 2008, Borrell et al., 2002). One of the proposed mechanisms for the effects includes disruption of cellular positioning secondary to altered cell migration (Koontz et al., 2008, Yan et al., 2004).

To this respect, primary cerebella cultures are a helpful *in vitro* model in the assessment of neuronal migration because they allow to getting valuable information on the role of interaction among all cell types present in the cerebellum, and also it is possible to evaluate migration of individual cells through the time. With this purpose, cells of the primary culture from mice at postnatal day (PND) 4th, time in which neuronal migration in the cerebellum is active are obtained; an additional advantage of the use of primary culture from cerebella of developing mice is that a high number of granule neurons as well as glia cells (astrocytes and microglia cells) is obtained.

In fact, migration of individual granule neurons may be followed by using cell culture plated on dishes with a CELLocate coverslip, etched with a 55-µm grid (Eppendorf, Hamburg, Germany), and coated with poly-D-lysine; because the coverslip is labeled at regular intersections with an alphanumerical system, making possible to identify the same location for tracking the same cells repeatedly in order to record their migration and to quantify the cell movement; tracking is accomplished by photodocumentation of coverslip using a microscope with phase-contrast optics, images are digitally recorded and the migration is quantified for each neuron, using the appropriated software (Sass et al., 2001). Using this model to assess neuronal migration, Sass et al. (2001) showed that LPS did not increase the number of neurons in movement, but increased the total distance of movement versus control cultures, and at the same time, in LPS-stimulated cultures showed an increase of TNF-α secretion and its mRNA expression.

After that, considering that data on gene expression profile is a useful approach to find genes and encoded proteins potentially involved in molecular mechanisms by which environmental factors affect neuronal migration, our group evaluated the effect of LPS, on the primary cerebella cultures from neonatal mice, in which previously was demonstrated that LPS alters migration and gene expression of some inflammatory cytokines (Sass et al., 2001). Using this approach, we found that more than one hundred genes were significantly modified by LPS, and the high proportion of these changes consisted in upregulation or induction of gene expression (Calderon-Aranda et al., unpublished data). Significant modification was considered when a relative change of >2-fold was found; we choose this cutoff since we considered that changes in gene expression from subtle to strong are relevant to a complex process like neuronal migration and for this reason we focus not only in one but in several potential targets, which must be at least, partially involved in regulation of neuronal migration. So, we think that there is no need to have strong changes in gene expression to be involved in the deregulation of neuronal migration. Interestingly LPS-upregulated genes included genes for several chemokines (e.g.*CCL2*, *CCL3*, *CCXL10*, *CXCL12*) that in different systems have been involved in neuronal migration and intracellular communication. In fact, these chemokines as well as their receptors have been located differentially in microglia and neurons in brain areas that included cortex and cerebellum (Bajetto et al., 2001), where it has been suggested they have a relevant function in brain development and can interact with neurotransmitter system (Adler et al., 2006). As it was expected, other genes modified by LPS included genes encoding proteins involved in activation pathways associated to TLR4, and then inflammatory response, for regulation of transduction and transcription factors (e.g. *IκBα*), cytokines, and cytokine's receptor (e.g. *TNF* and *oncostatin*), as well as genes involved in cellular signaling like several *Ras homology gene family* members, and cytoskeleton and extracellular matrix genes, some of which are known to be relevant to cell migration. This information has been extremely useful to address our research studies on mechanisms of LPS on neuronal migration dysfunction.

Based on information provided by gene expression study of cultures exposed to LPS, our group has been studying the role of several selected molecules using primary cerebella culture and neuroblastoma cell lines as models for mechanistic studies. In primary cerebella cultures, we recently showed that nitric oxide ($NO^{\bullet-}$) produced by inducible nitric oxide synthase (iNOS) through NF-κB signaling pathway is involved in neuronal migration induced by LPS and that in production of $NO^{\bullet-}$ through iNOS are involved TNF-α and IL- 1β induced by LPS in the primary culture (Arias-Salvatierra et al., 2011). Taken together, these results suggest that during development, pro-inflammatory stimulus are capable of modifying neuronal migration through intracellular signaling associated with receptors to PAMP (pathogen associated molecular pattern) like LPS or other molecules recognized by pattern recognition receptors.

Methylmercury

Evidence that MeHg disrupts neuronal migration came from the disasters occurred in Minamata Bay (Japan) and Iraq, where the population was exposed to low doses of MeHg during long or short periods to very high doses, respectively. As a result of these events, children exposed *in utero* to MeHg developed severe alterations in CNS as was described in section 2. As result of these disasters were obtained robust evidence of severe effects of MeHg on neuronal migration *in vivo* in humans and also in rodent models exposed *in utero* at different doses (Monnet-Tschudi et al., 1996; Kakita et al., 2002), and *in vitro* using cell lines and organotypic cultures, supporting that the effects of MeHg in neurodevelopment are associated, at least in part with inhibition of neuronal migration (Rice and Gilbert, 1990; Burbacher et al., 1990; Choi, 1986; Choi et al., 1978; Takeuchi, 1977). The inhibition of neuronal migration can be induced by relatively low doses or concentrations of MeHg *in vivo* and *in vitro; in vitro*, MeHg (0.1–10 μM) inhibits the movement of murine cerebellar neurons in culture (Jacobs et al., 1986; Kunimoto and Suzuki, 1997; Monnet-Tschudi et al., 1996; Sass et al., 2001). Sass et al., (2001) in primary cerebellum culture obtained from mice at PND 4th, found that 100 nM of MeHg in non-stimulated cells did not affect neuronal migration neither expression of inflammatory cytokines TNF-α and IL- 1β. Interestingly, when cultures were co-exposed to LPS and MeHg, neuronal migration was modified, compared with non-treated cultures (Sass et al., 2001). Our group evaluated the profile of gene expression in primary cerebellar culture exposed to 100 nM of MeHg (obtained as described Sass et al., 2001 and Arias-Salvatierra et al., 2011), with the purpose of generating information on target genes (and encoded proteins) affected by MeHg and potentially involved in neuronal migration process. In cultures treated with MeHg, we identified more than 100 genes modified; from these, the main proportion corresponded to non expressed genes in control cultures and approximately 30% of genes were upregulated (Calderon-Aranda et al., unpublished data). Among genes modified by MeHg, are included genes for proteins involved in NFκB regulation (e.g. *IκBα*), signal transducer and activator of transcription 3 (Stat3), cytoskeleton and extracellular matrix, some of which are involved in cell migration. These results are relevant since Stat3 is a crucial component for intracellular pathways associated to several cytokine receptors and growth factors (Heinrich et al., 2003). In fact, studies in knockout mice had demonstrated that Stat3 is implicated in migration of cells such as keratinocytes (Sano et al., 1999), and it was reported that independent of its phosphorylation state, Stat3 has a role in cell migration (Ng et al., 2006, Iancu et al., 2001). In preliminary studies our group has shown that MeHg targeted Stat3 phosphorylation in the same cell model used for gene expression studies (Calderon-Aranda et al., 2005).

Effect of MeHg on genes that encoded chemokines, cytoskeleton and extracellular matrix or growth factors proteins are also important since it is known they are involved in cell migration. Among these genes are included the *chemokines (C-X-C motif) ligand 5* that is known to play a role in brain development (Cartier et al., 2005); *thrombospondin 2*, a member of a family of extracellular proteins that participate in cell-to-cell and cell-to-matrix communication (Adams, 2001) and is a synaptogenic protein within a specific window of CNS development (Christopherson et al., 2005); *coronin*, a member of conserved F-actin binding proteins that are involved in signal transduction, transcriptional regulation, remodeling processes fibroblast migration and chemotaxis (Uetrecht and Bear, 2006);

*platelet-derived growth facto*r (PDGF), a potent growth factor that promotes the survival, migration, and differentiation of subventricular zone cells into myelin-forming oligodendrocytes (Lachapelle et al., 2002).

Based on gene expression data, the role of some candidate molecules have been evaluated by our group in the primary cerebellum culture described (Sass et al., 2001, Arias-Salvatierra et al., 2011), however to get more specific information on the role of a particular pathway, we are using cell lines well characterized in which the effect on migration is evaluated by the scratch method (Liang et al., 2007). Recently, our group found that neuronal migration is diminished by 100 nM of MeHg in primary cerebella cultures and this effect is related to NFkB activation and expression of iNOS (Arias-Salvatierra et al., submitted). Additionally, we have studied the MeHg effects on intracellular signaling induced by receptors for molecules involved in regulation of neuronal migration, as well as in regulation of cytoskeleton dynamics and assembly/disassembly of focal adhesions (FA); data from our group have shown that MeHg at non cytotoxic concentration interferes with activity of MAPK's and alters FA formation (manuscript in preparation) and currently, results obtained in primary cerebella cultures, and also in the neuroblastoma cell line SH-SY5Y, are currently in study in order to get more detailed information on the pathways affected by MeHg in cells induced to migrate.

Lead

Despite the extended knowledge about the effects of Pb in CNS, the molecular mechanisms are not yet completely defined. Here briefly, we will focus in the state of the art about the mechanisms of Pb neurotoxicity, like the signaling pathways altered and a succinct description of the normally used models to study the molecular mechanisms of neurotoxic effects of this metal. The Pb is capable to enter biological systems by food, water, air and soil (Toscano and Guilarte, 2005) and the exposure is very dangerous to humans, especially for children at developmental stages in which can result in permanent brain damage. Nowadays, the threshold of lead levels established by the Center for Disease Control (CDC) for cognitive deficits in children is 10 μg/dL (Landrigan, 2000) nevertheless, evidence has shown that PbB levels <10 μg/dL are associated with cognitive deficits in children (Canfield et al., 2003; Lanphear et al., 2005; Bellinger, 2008). Jusko et al. (2008) in a cohort study in children from 6 months to 6 years, shown that PbBconcentrations are inversely related to IQ scores; that is, the children's intellectual functioning at 6 years of age is impaired by PbB concentrations <10 μg/dL. Hereby, it has concluded that the effects of Pb are in function of the developmental stage when exposure occurs and the internal dose. It has been proposed that the general mechanisms of Pb toxicity (Bradberry and Vale, 2007) are related to: a) the formation of complexes with some chemical groups (-SH, NH_2 and –COOH), thereby disrupting the functionality of enzymes and proteins, b) the substitution of divalent ions (Ca^{2+}) altering Ca^{2+} mediated cellular processes, and c) oxidative stress induced by Pb produced by generation of reactive oxygen and nitrogen species (ROS and NOS) and decreasing antioxidant capability like GSH, metallothionein, catalase superoxide dismutase (SOD), glutathione peroxides, hemoxigenase, among others (Ahamed and Siddiqui, 2007; Yee y Choi, 1996). Actually, Pb binds to GSH as well as to the enzymes that have functional –SH groups like the δ-aminolaevulinate dehydrase (ALAD) and GSH reductase (Sandhir et al., 1994), and induces lipid peroxidation (Antonio-Garcia et al. 2008).

The majority of studies that assess the effect of Pb exposure in cognitive and behavioral function have been performed in non-human primates and rodents (Cory-Slechta, 1995, Rice, 1984, 1990). In the early 90's it was demonstrated that Pb specifically blocked the ionotropic glutamate receptor, the N-Methyl-D-Aspartate Receptor (NMDAR), having a direct effect in spatial learning (Alkondon et al., 1990; Guilarte and Miceli, 1992), a function hippocampal and NMDAR-dependent (Morris et al., 1982, 1986) The learning alteration is related to disruption of hippocampal long-term potentiation (LTP) due to exposure to Pb through development, and Pb effects could be long lasting in function of the stage of exposure and dosing. LTP is a synaptic process NMDAR-dependent widely investigated and used as a model of learning and memory and defined as a long-lasting increase in synaptic efficacy subsequent to short periods of stimulation of specific synapses (Bliss and Lomo, 1973). At this respect, Guilarte and McGlothan (1998) using *in situ* hybridization, found that exposure to Pb during development alters NMDAR subunit expression, increases NR1 subunit expression in hippocampus and diminishes NR2A subunit at 14 and 21 PND. Interestingly, the mRNA expression of subunits NR1, NR2A-C was not affected in cerebellum of the treated rats, which indicates a selectivity of Pb effects on hippocampus. These results suggest that Pb exposure can stimulate changes in the composition of NMDAR that could lead to alterations in the threshold of physiological events like LTP, reaching detrimental effects in learning and memory of exposed individuals. Alteration in the expression pattern of these receptors could also modify intracellular signaling and the expression genes of some neural factors that are not solely implied in learning processes, but also in brain development and maturation (Nihei and Guilarte, 2001). Currently, Pb exposure is well established that disrupts NMDAR-dependent processes like hippocampal LTP (Gilbert and Mack, 1998, Gilbert et al., 1996, Altmann et al., 1993), spatial learning (Kuhlmann et al., 1997, Nihei et al., 2000) and the development of the barrel field cortex in animal models (Wilson et al., 2000). These *in vivo* studies provide evidence to support the detrimental effect of exposure to Pb on cerebral function during brain development.

It is important to underline moreover that Pb affects learning and memory, and the obvious role of NMDAR in these alterations. Recent studies *in vitro* have been focused in the role of neural cell adhesion molecule (NCAM) as a possible molecular target of Pb due to its role in cell adhesion, neuronal migration, synaptic plasticity, learning and memory; Hu et al. (2011) reported in hippocampal neurons obtained from embryonic Wistar pups (embryonic day 18) that exposure to Pb reduced the survival rate of cultures and delayed the neurite outgrowth in a wide range of Pb concentrations (1 nM-10 μM). This information provides a new field of investigation about the specific molecular mechanisms besides the already known in the broad spectrum of Pb induced alterations in CNS.

Alcohol

There are different points in the development of CNS that can be altered by alcohol exposure, these can be seen in children born to chronically alcoholic mother, and these symptoms are known as the FAS. Children with FAS present a variety of neural and extraneural abnormalities and when do not exhibit completely the symptoms of FAS, may present dimorphic, cognitive and behavioral effects that are known as fetal alcohol effects (Kumada et al., 2007). At molecular level, alcohol is capable to alter signaling pathways, voltage-gated ion channels, neurotransmitter receptors and membrane fluidization (West et al., 1994; Costa et al., 2004; Olney, 2004), and also causes disruption of neuronal migration. The commonly used model to study neuronal migration alterations by alcohol exposure is the cerebellum of early postnatal mice due to vulnerability of cerebellum at this stage of development (comparable to cerebellum of humans in the third gestational trimester) (Sakata-Haga et al., 2001; Dikranian et al., 2005). The low number of granule cells in the internal granular layer in treated animals, suggested that alcohol interferes with the migration of granule cells from external to internal granule layer (Borges and Lewis, 1983). However, the specific mechanism by which alcohol causes these disturbances are not well understood and it has been recently proposed that Ca^{2+} and nucleotide signaling are the principal targets of alcohol (Kumada et al., 2007). Low alcohol concentrations can alter the functions of Ca^{2+} channels (Walter and Messing, 1999), suggesting that alcohol may alter the intracellular Ca^{2+} in neurons at different levels, including alterations in PKC signaling pathway that could directly disrupt cell migration, since this kinase is implicated in actin polymerization (Tasken and Aandahl, 2004). Alcohol exposure also affects cAMP and cGMP levels, increasing and decreasing their levels respectively, and thereby stimulating the pathways associated to them. The participation of these molecules was corroborated *in vivo*, showing that pharmacological manipulation of the signaling pathways associated to Ca^{2+}, cAMP and cGMP reduced the adverse effects of alcohol exposure on neuronal migration (reviewed in Kumada et al., 2007). Recent studies by Aronne et al. (2011) suggest that alterations in corticogenesis observed in alcohol treated animals could be result of different mechanisms, including: 1) impaired neuroblast generation, 2) retarded neuroblast migration, or 3) impairment of the scaffold through which cells migrate. From these studies, it was suggested that molecules like vimentin, nestin, S-100b, Pax6 and doublecortin, deserve attention as possible molecular targets of alcohol toxicity (Aronne et al., 2011).

Conclusion

In this chapter we present the current evidences on neurological disorders associated to environmental factors, as well as the knowledge on molecular mediators of deregulation of neuronal migration, a neurodevelopmental alteration associated to cerebral palsy, schizophrenia, ASD, and ADHD, among other NDDs. The purpose was to contribute to a better understanding of the etiology, physiopathology and environmental risk factors for neurodevelopment disorders in which neuronal migration disruption is involved.

As it was described, a group of NDDs shares the etiopathogenesis, involving the interaction of genetic influences and environmental risks, such as pathogens operating on brain maturational processes, neurotoxic metals like Hg and Pb, alcohol, among other environmental factors. Epidemiological studies have proved enough evidence of the role of environmental factors on NDDs.

In the last two decades experimental evidence suggests that the neuronal migration disorder is one of the molecular mechanisms associated with NDDs. However, in spite of the numerous efforts, molecular mechanisms involved in neuropathogenesis are not completely known, due to the enormous complexity of the processes potentially affected. Nevertheless growing evidence supports the involvement of inflammation mediators, dysfunction of protein by metal through binding to thiols, oxidant damage, alteration of Ca^{2+}, deregulation of cell signaling, neuronal migration, impairment of neurotransmission, among others, have an important role in the pathogenesis of NDDs.

The fact is that one of the mechanisms that have been involved in the neuronal migration deregulation is associated to induction of inflammation, directly generated in the nervous system in response to infection by neurotropic pathogens or chemical compounds (central inflammatory process), or as consequence of an effect at peripheral level in pregnant mothers that may cause transplacental passage of pathogen to the fetus; this mechanism is not surprising because IL-1β, IL-6, and TNF-α are augmented in the pregnant mothers after maternal infection. Pro-inflammatory cytokines are known to regulate normal brain development and have also been implicated in abnormal corticogenesis. The role of inflammation in NDDs has been supported by studies using *E. coli* cell wall endotoxin LPS, which causes disruption of sensorimotor gating in the offspring, suggest that direct stimulation of cytokine production by infections or immunogenic agents, as well as environmental factors (heavy metals like Hg and Pb) acting as risk cofactors for infection or autoimmunity due to infectious agents. All together these data suggest that they may be considered seriously as etiological cofactors for NDDs.

Besides the study of genetic factors associated with diseases related to NDDs, more studies are needed to establish the molecular mechanisms of toxicity associated with the risk of NDDs in response to exposure to environmental factors. In fact, with the knowledge of molecular mechanisms related to these disorders, a real challenge to researchers and physicians is the identification and validation of biomarkers of prenatal early occurrence of neurodevelopmental effect in response to environmental insults. With this purpose, the study of alteration of the normal gene expression and proteins involved in glial and neuronal genesis, cell proliferation, migration, axonal outgrowth, synaptogenesis, and apoptosis may be extremely useful to rationally choose of sensitive biomarkers of early effects during the critical time windows affecting the regulation of neurodevelopment.

Finally, we can conclude that an important strategy to decrease impact of environmental factors in neurodevelopment, as well as incidence and severity of disorders associated to neurodevelopment disorders, is to identify high risk population exposed to environmental factors. Taken together all these strategies, should allow the implementation of intervention to decrease the exposure to these environmental pollutants, as well as diagnostic screening using really early risk markers of effect in order to perform therapeutic strategies to limit the impact of neurodevelopment disruption.

Studies from our group were funded by grants from Cure Autism Now, Heinz Family Foundations and by the Mexican Council for Science and Technology (Conacyt 46297-M).

References

Adler, M. W. Geller, E. B. Chen, X. Rogers, T. J. (2006). Viewing chemokines as a third major system of communication in the brain. *AAPS J, 7,* E865-E870.

Ahamed, M. & Siddiqui, M. K. (2007). Low level lead exposure and oxidative stress: current opinions. *Clin Chim Acta, 383*, 57-64.

Akbarian, S. & Huang, H-S. (2006). Molecular and cellular mechanisms of altered GAD1/GAD67 expression in schizophrenia and related disorders. *Brain Res Rev, 52,* 293-304.

Akbarian, S. Bunney, W. E. Potkin, S. G. Wigal, S. B. Hagman, J. O. Sandman, C. A. Jones, E. G. (1993). Altered distribution of nicotinamide-adenine dinucleotide phosphate-diaphorase cells in frontal lobe of schizophrenic implies disturbances of cortical development. *Arch Gen Psychiatry, 50,* 169-177.

Alkondon, M., Costa, A. C. Radhakrishnan, V. Aronstam, R. S. Albuquerque E. X. (1990). Selective blockade of NMDA-activated channel currents may be implicated in learning deficits caused by lead. *FEBS Lett, 261,* 124-130.

Altmann, L. Weinsberg, F. Sveinsson, K. Lilienthal, H. Wiegand, H. Winneke, G. (1993). Impairment of long-term potentiation and learning following chronic lead exposure. *Toxicol Lett, 66,* 105-112.

Andreeva, A. V. & Kutuzov, M. A. (2010). Cadherin 13 in cancer. *Genes Chromosomes Cancer, 49,* 775-790.

Andrews, N. Miller, E. Grant, A. Stowe, J. Osborne, V. Taylor, B. (2004). Thimerosal exposure in infants and developmental disorders: a retrospective cohort study in the United Kingdom does not support a causal association. *Pediatrics, 114,* 585-591.

Antonio-Garcia, M. T. & Masso-Gonzalez, E. L. (2008). Toxic effects of perinatal lead exposure on the brain of rats: involvement of oxidative stress and the beneficial role of antioxidants. *Food Chem Toxicol, 46*, 2089-2095.

Archibald, S. L. Fennema-Notestine, C. Gamst, A. Riley, E. P. Mattson, S. N. Jernigan, T. L. (2001). Brain dysmorphology in individuals with severe prenatal alcohol exposure. *Dev Med Child Neurol, 43,* 148-154.

Arias-Salvatierra, D. Acosta-Saavedra, L. C. Silbergeld E. K. Calderon-Aranda E. S. Methylmercury inhibits the NF-κB signaling pathway and inducible nitric oxide production in primary cerebellar cultures: a molecular mechanism related to effect on neuronal migration of granule neurons, submitted.

Arias-Salvatierra, D. Silbergeld E. K. Acosta-Saavedra, L. C. Calderon-Aranda E. S. (2011) Role of nitric oxide produced by iNOS through NF-κB pathway in migration of cerebellar granule neurons induced by Lipopolysaccharide. *Cell Signal, 23*, 425-435.

Arnold, S. E. Hyman, B. T. Van Hoesen, G. W. Damasio, A. R. (1991). Some cytoarchitectural abnormalities of the entorhinal cortex in schizophrenia. *Arch Gen Psychiatry, 48,* 625-632.

Arnold, S.E. (1999). Neurodevelopmental abnormalities in schizophrenia: insights from neuropathology. *Dev. Psychopathol, 11*, 439-56.

Aronne, M. P. Guadagnoli, T. Fontanet, P. Evrard, S. G. Brusco, A. (2011). Effects of prenatal ethanol exposure on rat brain radial glia and neuroblast migration. *Exp Neurol,* 229, 364-371.

Arpino, C. Marzio, M. D'Argenzio, L. Longo, B. Curatolo, P. (2005). Exanthematic diseases during pregnancy and attention-deficit/hyperactivity disorder (ADHD). *Eur J Paediatr Neurol, 9,* 363-365.

Atladóttir, H. O. Thorsen, P. Østergaard, L. Schendel, D. E. Lemcke, S. Abdallah, M. Parner, E. T. (2010). Maternal infection requiring hospitalization during pregnancy and autism spectrum disorders. *J Autism Dev Disord, 40,* 1423-1430.

Bajetto, A. Bonavia, R. Barbero, S. Florio, T. Schettini, G. (2001). Chemokines and their receptors in the central nervous system. *Front Neuroendocrinol, 22*, 147-184.

Banaschewski, T. Becker, K. Scherag, S. Barbara, F. Coghill, D. (2010). Molecular genetics of attention-deficit/hyperactivity disorder: an overview. *Eur Child Adolesc Psychiatry, 19,* 237-257.

Bell, S.H. Stade, B. Reynolds, J.N. Rasmussen, C. Andrew, G. Hwang, P.A. Carlen, P. L. (2010). The remarkably high prevalence of epilepsy and seizure history in fetal alcohol spectrum disorder. *Alcohol Clin Exp Res, 34,* 1084-1089.

Bellinger, D. C. (2008). Very low lead exposures and children's neurodevelopment. *Curr Opin Pediatr, 20*, 172-177

Bishop, D. & Rutter, M. (2008). Neurodevelopmental disorders: conceptual approaches. In: M. Rutter, D. Bishop, D. Pine, S. Scott, J. Stevenson, E. Taylor. & A. Thapar editors. Rutter's Child and Adolescent Psychiatry Oxford: Blackwell pp 32-41.

Bliss, T. V. & Lomo, T. (1973). Long-lasting potentiation of synaptic transmission in the dentate area of the anaesthetized rabbit following stimulation of the perforant path. *J Physiol, 232*, 331-356.

Borges, S. & Lewis, P. D. (1983). Effects of ethanol on postnatal cell acquisition in the rat cerebellum. *Brain Res, 271*, 388-391.

Borrell, J. Vela, J.M. Arévalo-Martin, A. Molina-Holgado, E. Guaza, C. (2002). Prenatal immune challenge disrupts sensorimotor gating in adult rats. Implications for the etiopathogenesis of schizophrenia. *Neuropsychopharmacol, 26,* 204-15.

Bradberry, S. Vale, A. (2007). Lead. *Medicine, 32*, 627-628.

Braun, J. M. Kahn, R. S. Froehlich, T., Auinger, P., Lanphear, B. P. (2006). Exposures to environmental toxicants and atteintion deficit hyperactivity disorder in U.S. children. *Environ Health Perspect, 114,* 1904-1909.

Brown, A. S. Deicken, R. F. Vinogradov, S. Kremen, W.S. Poole, J. H. Penner, J. D. Kochetkova, A. Kern, D. Schaefer, C. A. (2009a). Prenatal infection and cavum septum pellucidum in adult schizophrenia. *Schizophr Res, 108,* 285-287.

Brown, A. S. Hooton, J. Schaefer, C.A. Zhang, H. Petkova, E. Babulas, V. Perrin, M. Gorman, J. M. Susser, E. S. (2004). Elevated maternal interleukin-8 levels and risk of schizophrenia in adult offspring. *Am J Psychiatry, 161,* 889-895.

Brown, A. S. Vinogradov, S. Kremer, W. S. Poole, J.H. Deiken, R. F. Penner, J. D. McKeague, I. W. Kochetkova, A. Kern, D. Schaefer, C. A. (2009b). Prenatal exposure to maternal infection and executive dysfunction in adult schizophrenia. *Am J. Psychiatry, 166,* 683-690.

Brown, A.S. & Susser, E.S. (2002). In utero infection and adult schizophrenia. *Ment Retard Dev Disabil Res Rev, 8,* 51-57.

Burbacher, T. M. Rodier, P. M. Weiss, B. (1990). Methylmercury developmental neurotoxicity: A comparison of effects in humans and animals. *Neurotoxicol Teratol, 12,* 191–202.

Calderon-Aranda, E. C. Acosta-Saavedra, L. C. Silbergeld, E.K. (2005) Methyl mercury (MeHg) alters IL-6 signal transduction in mouse cerebellar cells in culture. *Toxicol Sci, 84,* 627.

Campbell, D. B. D'Oronzio, R. Garbett, K. Ebert, P. J. Mirnics, K. Levitt, P. Persico, A. M. (2007). Disruption of cerebral cortex MET signaling in autism spectrum disorder. *Ann Neurol, 62,* 243-250.

Campbell, D. B. Li, C. Sutcliffe, J. S. Persico, A. M. Levitt, P. (2008). Genetic evidence implicating multiple genes in the MET receptor tyrosine kinase pathway in autism spectrum disorder. *Autism Res, 1,* 159-168.

Campbell, D. B. Warren, D. Sutcliffe, J. S. Lee, E. B. Levitt, P. (2010). Association of MET with social and communication phenotypes in individuals with autism spectrum disorder. *Am J Med Genet B Neuropsychiatr Genet, 153B,* 438-446.

Canfield, R. L. Henderson. Jr. C. R. Cory-Slechta, D. A. Cox, C. Jusko, T. A. Lanphear, B. P. (2003). Intellectual impairment in children with blood lead concentrations below 10 microg per deciliter. *N Engl J Med, 348,* 1517–1526.

Cartier, L. Hartley, O. Dubois-Dauphin, M. Krause, K. H. (2005). Chemokine receptors in the central nervous system: role in brain inflammation and neurodegenerative diseases. *Brain Res Rev 48*, 16-42.

CDC (Centers for Disease Control and Prevention). Preventing Lead Poisoning in Young Children: A Statement by the Centers for Disease Control. Centers for Disease Control, Atlanta, GA. 1991.

Chakrabarti, S. & Fombone, E. (2005). Pervasive developmental disorders in preschool children: confirmation of high prevalence. *Am J Psychiatry, 162*, 1133-1141.

Chang, B. S. Ly, J. Appignani, B. Bodell, A. Apse, K. A. Ravenscroft, R. S. Sheen, V. L. Doherty, M.J. Hackney, D. B. O'Connor, M. Galaburda, A. M. Walsh, C. A. (2005). Reading impairment in the neuronal migration disorder of periventricular nodular heterotopias. *Neurology, 64,* 799-803.

Chevassus-au-Lous, N. & Represa, A. (1999). The right neuron at the wrong place: biology of heterotopic neurons in cortical neuronal migration disorders, with special reference to associated pathologies. *CMLS, 55,* 1206-1215.

Choi, B. H. (1986). Methylmercury poisoning of the developing nervous system: I. Pattern of neuronal migration in the cerebral cortex. *Neurotoxicol, 7*, 591-600.

Choi, B. H. Lapham, L. W. Amin-Zaki, L. Saleem, T. (1978). Abnormal neuronal migration, deranged cerebral cortical organization, and diffuse white matter astrocytosis of human fetal brain: a major effect of methylmercury poisoning in utero. *J Neuropathol Exp Neurol, 37*, 719-733.

Christopherson, K. S. Ullian, E. M. Stokes, C. C. Mullowney, C. E. Hell, J.W. Agah, A. Lawler, J. Mosher, D.F. Bornstein, P. Barres, B.A. (2005). Thrombospondins are astrocyte-secreted proteins that promote CNS synaptogenesis. *Cell, 11*, 421-433.

Cirulli, F. Laviola, G. Ricceri, L. (2009). Risk factors for mental health: translational models from behavioral neuroscience. *Neurosci Biobehav Rev, 33,* 493-497.

Clarren, S. K. Alvord, E. C. Sumi, S. M. Streissguth, A. P. Smith, D. W. (1978). Brain malformations related to prenatal exposure to ethanol. *J Pediatr, 92,* 64-67.

Clegg, D. O. Wingerd, K. L. Hikita, S. T., Tolhurst, E. C. (2003). Integrins in the development, function and dysfunction of the nervous system *Frontiers in Biosci, 8*, 723-750.

Cohen, J. T. Bellinger, D. C. Shaywitz, B. A. (2005). A quantitative analysis of prenatal Methylmercury exposure and cognitive development. *Am. J. Prev. Med, 29,* 353-365.

Cohly, H. H. & Panja, A. (2005). Immunological findings in autism. *Int Rev Neurobiol, 71,* 317-341.

Cory-Slechta D. A. (1995). Relationships between lead-induced learning impairments and changes in dopaminergic, cholinergic, and glutamatergic neurotransmitter system functions. *Annu Rev Pharmacol Toxicol, 35*, 391-415.

Costa, L .G. Vitalone, A. Guizzetti, M. (2004). Signal transduction mechanisms involved in the antiproliferative effects of ethanol in glial cells. *Toxicol Lett, 149*, 67-73.

Dikranian, K. Qin, Y.Q. Labruyere, J. Nemmers, B. Olney, J.W. (2005). Ethanol-induced neuroapoptosis in the developing rodent cerebellum and related brain stem structures. *Dev Brain Res, 155*, 1-13.

Faraone, S. V. Perlis, R. H. Doyle, A. E. Smoller, J. W. Goralnick, J. J. Holmgren, M. A. Sklar, P. (2005). Molecular genetics of attention deficit/hyperactivity disorder. *Biol Psychiatry, 57,* 1313-1323.

Fatemi, S. H. Earle, J. A. Menomy, T. (2000). Reduction in reelin immunoreactivity in hippocampus of subjects with schizophrenia, biporla disorder and major depression. *Mol. Psychatry, 5,* 654-663.

Fatemi, S. H. Kroll, J. L. Stary, J. M. (2001). Altered levels of reelin and its isoforms in schizophrenia and mood disorders. *Neuroreport, 12,* 3209-3215.

Fatemi, S. H. Snow, A. V. Stary, J. M. Araghi-Niknarn, M. Reutiman, T. J. Lee, S. Brooks, A. I. Pearce, D. A. (2005). Reelin signaling is impaired in autism. *Biol Psychiatry, 57,* 777-787.

Fatemi, S. H. Stary, J. M. Egan, E. A. (2002). Reduced blood levels of reelin as a vulnerability factor in pathophysiology of autistic disorder. *Cell Mol Neurobiol, 22,* 139-152.

Frahm, C. & Draguhn, A. (2001). GAD and GABA transporter (GAT-1) mRNA expression in the developing rat hippocampus. *Brain Res Dev Brain Res, 132,* 1-13.

Franke, B. Neale, B. M. Faraone, S. V. (2009). Genome-wide association studies in ADHD. *Hum Genet, 126,* 13-50.

Froehlich, T. E., Lanphear, B. P., Auinger, P., Hornung, R., Epstein, J. N., Braun, J., Kahn, R. S. (2009). Association of tobacco and lead exposures with attention deficit/hyperactivity disorder. *Pediatrics, 124,* 1054-1063.

Garay, P.A. & McAllister, A.K. (2010). Novel roles for immune molecules in neural development: implication for neurodevelopmental disorders. *Frontiers in synaptic neuroscience, 2,* 1-16.

Geier, D. A. & Geier, M. R. (2005). A two-phased population epidemiological study on the safety of thimerosal-containing vaccines: a follow-up analysis. *Med Sci Monitor, 11,* CR160-CR170.

Geier, D. A. Kern, J. K. Geier, M. R. (2009). A prospective study of prenatal mercury exposure from maternal dental amalgams and autism severity. *Acta Neurobiol Exp, 69,* 189-197.

Gilbert, M. E. & Mack, C. M. (1998). Chronic lead exposure accelerates decay of long-term potentiation in rat dentate gyrus in vivo. *Brain Res, 789*, 139-149.

Gilbert, M. E. Mack, C. M. Lasley, S. M. (1996) Chronic developmental lead exposure increases the threshold for long-term potentiation in rat dentate gyrus in vivo. *Brain Res, 736*, 118-124.

Goncalves, A S. & Appelberg, R. (2002). The involvement of the chemokine receptor CXCR2 in neutrophil recruitment in LPS-induced inflammation and in Mycobacterium avium infection. *Scand J Immunol, 55*, 585-591.

Gressens, P. (2005). Neuronal Migration Disorders. *J.Child.Neurol, 20,* 968-971.

Grigorenko, E. L. Wood, F.B. Meyer, M. S. Hart, L.A. Speed, W. C. Shuster, A. Pauls, D. L. (1997). Susceptibility loci for distinct components of developmental dyslexia on chromosomes 6 and 15. *Am J Hum Genet, 60,* 27-39.

Guerrini, R. & Barba, C. (2010). Malformations of cortical development and aberrant cortical netxorks: epileptogenesis and functional organization. *J Clin Neurophysiol, 27,* 372-379.

Guerrini, R. Dobyns, W. B. Barkovich, A. J. (2008). Abnormal development of the human cerebral cortex: genetics, functional consequences and treatment options. *Trends Neurosci, 31*, 154-162.

Guilarte, T. R. & McGlothan, J. L. (1998). Hippocampal NMDA receptor mRNA undergoes subunit specific changes during developmental lead exposure. *Brain Res, 790*, 98-107.

Guilarte, T. R. & Miceli, R. C. (1992). Age-dependent effects of lead on $^{[3H]}$MK-801 binding to the NMDA receptor-gated ionophore: in vitro and in vivo studies. *Neurosci Lett, 148,* 27-30.

Ha, M. Kwon, H. J. Lim, M. H. Jee, Y. K. Hong, Y. C., Leem, J. H. Sakong, J. Bae, J. M. Hong, S. J. Roh, Y. M. Jo, S. J. (2009). Low blood levels of lead and mercury and symptoms of attention deficit hyperactivity in children: a report of the children's health and environment research (CHEER). *Neurotoxicol, 30,* 31-36.

Happé, F. Ronald, A. Plomin, R. (2006). Time to give up on a single explanation for autism. *Nat Neurosci, 9,* 1218-1220.

Heinrich, P. C. Behrmann, I. Haan, S. Hermanns, H. M. Müller-Newen, G. Schaper, F. (2003). Principles of interleukin (IL)-6-type cytokine signaling and its regulation. *Biochem J, 374*, 1-20.

Holmes, P. James, K. A. Levy, L. S. (2009). Is low-level environmental mercury exposure of concern to human helath?. *Sci. Total Environ, 408,* 171-82.

Hu, Q. Fu, H. Song, H. Ren, T. Li, L. Ye, L. Liu, T. Dong, S. (2011). Low-level lead exposure attenuates the expression of three major isoforms of neural cell adhesion molecule. *Neurotoxicol, 32*, 255-260.

Iancu, C. Mistry, S. J. Arkin, S. Wallenstein, S. Atweh, G. F. (2001) Effects of stathmin inhibition on the mitotic spindle. *J Cell Sci, 114*, 909-916.

Jacobs, A. J. Maniscalco, W. M. Finkelstein, J. N. (1986). Effects of methylmercuric cloride, cycloheximide, and colchicine on the re-aggregation of dissociated mouse cerebellar cells. *Toxicol Appl Pharmacol, 86*, 362-371.

Jakob, H. & Beckmann, H. (1986). Prenatal developmental disturbances in the limbic allocortex in schizophrenics. *J Neural Transm, 65,* 303-326.

Jusko, T. A. Henderson, C. R.Jr. Lanphear, B .P. Cory-Slechta, D. A. Parsons, P. Canfield, R. L. (2008). Blood Lead concentrations < 10μg/dL and child intelligence at 6 years of age. *Environ Health Perspec, 116*, 243-248.

Kakita, A. Inenaga, C. Sakamoto, M. Takahashi, H. (2002). Neuronal migration disturbance and consequent cytoarchitecture in the cerebral cortex following transplacental administration of methylmercury. *Acta Neuropathol, 104*, 409-417.

Keller, F. & Persico, A. M. (2003). The neurobiological context of autism. *Mol Neurobiol, 28,* 1-22.

Koontz, T. Bralic, M. Tomac, J. Pernjak-Pugel, E. Bantug, G. Jonjic, S. Britt, W. J. (2008). Altered development of the brain after focal herpesvirus infection of the central nervous system. *J Exp Med, 205*, 423-435.

Kriegstein, A. R. (2005) Constructing circuits: neurogenesis and migration in the developing neocortex. *Epilepsia, 46*, 15-21.

Kuhlmann, A. C. McGlothan, J. L. Guilarte, T. R. (1997). Developmental lead exposure causes spatial learning deficits in adult rats. *Neurosci Lett, 233*, 101-104.

Kumada, T. Jiang, Y. Cameron, D. B. Komuro, H. (2007). How does alcohol impair neuronal migration? *J Neurosci Res, 85*, 465-470.

Kunimoto, M. & Suzuki, T. (1997). Migration of granule neurons in cerebellar organotypic cultures is impaired by methylmercury. *Neurosci Lett, 226*, 183-186.

Lachapelle, F. Avellana-Adalid, V. Nait-Oumesmar, B. Baron-Van Evercooren, A. (2002). Fibroblast growth factor-2 (FGF-2) and platelet-derived growth factor AB (PDGF AB) promote adult SVZ-derived oligodendrogenesis in vivo. *Mol Cell Neurosci, 20*, 390-403.

Landgren, M. Svensson, L. Strömland, K. Andersson, G.M. (2010). Prenatal alcohol exposure and neurodevelopmental disorders in children adopted from Eastern Europe. *Pediatrics, 125,* e1178-e1185.

Landrigan P. J. (2000). Pediatric lead poisoning: is there a threshold? *Public Health Rep, 115,* 530– 531.

Lanphear, B. P. Hornung, R. Khoury, J. Yolton, K. Baghurst, P. Bellinger, D. C. Canfield, R. L. Dietrich, K. N. Bornschein, R. Greene, T. Rothenberg, S. J. Needleman, H. L. Schnaas, L. Wasserman, G. Graziano, J. Roberts, R. (2005). Low-level environmental lead exposure and children's intellectual function: an international pooled analysis. *Environ Health Perspect, 113*, 894–899.

Laviola, G. Ognibene, E. Romano, E. Adriani, W. Keller, F. (2009). Gene-environment interaction during early development in the heterozygous reeler mouse: clues for modelling of major neurobehavioral sindromes. *Neurosci Biobehav Rev*, *33*, 560-572.

Lee, J. M. Tiong, J. Maddox, D. M. Condie, B. G. Wray, S. (2008). Temporal migration of gonadotrophin-releasing hormone-1 neurons is modified in GAD67 knockout mice. *J Neuroendocrinol, 20*, 93-103.

Lee, S. J. & Benveniste, E. N. (1999). Adhesion molecule expression and regulation on cells of the central nervous system. *J Neuroimmunol, 98*, 77-88.

Lewis, D. A. (2001). Retroviruses and the pathogenesis of schizophrenia. *Proc Natl Acad Sci USA, 94,* 4293-4294.

Liang, C. C, Park, A. Y. Guan, J. L. (2007). In vitro scratch assay: a convenient and inexpensive method for analysis of cell migration in vitro. *Nat Protoc, 2,* 329-33.

Maeda, N. Fukazawa, N. Ishii, M. (2010). Chondroitin sulfate proteoglycans in neural development and plasticity. *Front Biosci, 13,* 626-644.

Mann, J. R. & McDermott, S. (2010). Are maternal genitourinary infection and pre-eclampsia associated with ADHD in school aged children? *J Atten Disord, September 13,* 2010, doi: 10.1177/1087054710370566.

Maqueda, J. Ramirez, M. Lamas, M. Gutierrez, R. (2003). Glutamic acid decarboxylase (GAD)67, but not GAD65, is connstitutively expressed during development and transiently overexpressed by activity in the granule cells of the rat. *Neurosci Lett, 353,* 69-71.

Marín, O. & Rubenstein, J.L. (2003). Cell migration in the forebrain. *Annu Rev Neurosci, 23,* 441-483.

McGee, C. L. & Riley, E. P. (2006). Brain imaging and fetal alcohol spectrum disorders. *Ann Ist Super Sanita, 42,* 46-52.

Mei, L. & Xiong, W. C. (2008). Neuregulin 1 in neural development, synaptic plasticity and schizophrenia. *Nat Rev Neurosci, 6*, 437-452.

Mochida, G. H. & Walsh, C. (2004). Genetic basis of developmental malformations of the cerebral cortex. *Arch Neurol, 61,* 637-640.

Monnet-Tschudi, F. Zurich, M.G. Honegger, P. (1996). Comparison of the developmental effects of two mercury compounds on glial cells and neurons in aggregate cultures of rat telencephalon. *Brain Res, 741*, 52-59.

Morris, D. W. Robinson, L. Turic, D. Duke, M. Webb, V. Milham, C. Hopkin, E. Pound, K. Fernando, S. Easton, M. Hamshere, M. Williams, N. McGuffin, P. Stevenson, J. Krawczak, M. Owen, M.J. O'Donovan, M. C. Williams, J. (2000). Family-based association mapping provides evidence for a gene for reading disability on chromosome 15q. *Hum Molec Genet, 9,* 843-848.

Morris, R. G. Anderson, E. Lynch, G. S. Baudry, M. (1986). Selective impairment of learning and blockade of long-term potentiation by an N-methyl-d-aspartate receptor antagonist, AP5. *Nature, 319*, 774-776.

Morris, R. G. Garrud, P. Rawlins, J.N. O'Keefe, J. (1982). Place navigation impaired in rats with hippocampal lesions, *Nature, 297*, 681- 683.

Mortensen, P. B. NØrgaard-Pedersen, B. Waltoft, B.L, SØrensen, T.L. Hougaard, D. Yolken, R. H. (2007). Early infections of Toxoplasma gondii and the later development of schizophrenia. *Schizophr Bull, 33,* 741-744.

Mutter, J. Naumann, J. Schneider, R. Walach, H. Haley, B. (2005). Mercury and autism: accelerating evidence? *Neuro Endocrinol Lett, 26,* 439-446.

Ng, D. C. Lin, B. H. Lim, C. P. Huang, G. Zhang, T. Poli, V. Cao, X. (2006). Stat3 regulates microtubules by antagonizing the depolymerization activity of stathmin. *Cell Biol, 172*, 245-257.

NIH: National Institutes of Health Consensus Development Conference Statement: diagnosis and treatment of attention-deficit/hyperactivity disorder (ADHD). *J Am Acad Child Adolesc Psychiatry, 39*, 182-193.

Nihei, M. K, & Guilarte T. R. (2001). Molecular changes in glutamatergic synapses induced by Pb2+: association with deficits of LTP and spatial learning. *Neurotoxicol, 22,* 635-643.

Nihei, M. K. Desmond, N. L. McGlothan, J. L. Kuhlmann, A. C. Guilarte, T. R. (2000). N-methyl-d-aspartate receptor subunit changes are associated with lead-induced deficits of long-term potentiation and spatial learning. *Neuroscience, 99*, 233-242.

Nopoulos, P. Berg, S. Castellenos, F. X. Delgado, A. Andreasen, N. C. Rappaport, J. L. (2000). Developmental brain anomalies in children with attention deficit hyperactivity disorder. *J Child Neurol. 15,* 102-108.

NRC. National Research Council. 2000. Toxicological Effects of Methylmercury. Washington, DC. National Academy Press; 33-35.

O'Malley, K. D. & Nanson, J. (2002). Clinical implications of a link between fetal alcohol spectrum disorder and attention-deficit hyperactivity disorder. *Can J Psychiatry, 47,* 349-354.

Olney, J. W. (2004). Fetal alcohol syndrome at the cellular level. *Addict Biol, 9*, 137-149.

Opler, M. G. Brown, A. S., Graziano, J. Desai, M. Zheng, W. Schaefer, C. Factor-Litvak, P. Susser, E. S. (2004). Prenatal lead exposure, delta-aminolevulinic acid, and schizophrenia. *Environ Health Perspect, 112,* 548-552.

Pantazopoulos, H. Woo, T. U. W. Lim, M. P. Lange, N. Berretta, S. (2010). Extracellular matrix-glial abnormalities in the amygdale and entorhinal cortex of subjects diagnosed with schizophrenia. *Arch Gen Psychiatry, 67,* 155-166.

Polanczyk, G. Silva de Lima, M. Lessa-Horta, B. Biederman, J. Rohde, L. A. (2007). The worldwide prevalence of ADHD: A systematic review and metaregression analysis. *Am J Psychiatry, 164*, 942-948.

RACP. Royal Australasian College of Physicians. *Australian Guidelines on Attention Deficit Hyperactivity Disorder (ADHD)* National Health and Medical Research Council; 2009. [Accessed May 5th, 2011]. Available at: http://nhmrc.gov.au/_files_nhmrc/file/publications/synopses/adhd/NHMRC-draft-DHD-guidelines.pdf

Rakic, P., Hashimoto, T. K., Sarkisian, M. R. (2007). Genetic determinants of neuronal migration in the cerebral cortex. *Novartis Found Symp, 288,* 45-53.

Ramsay, M. (2010). Genetic and epigenetic insights into fetal alcohol spectrum disorders. *Genome Med, 2,* 27-34.

Rapoport, J. L., Addington, A. M., Frangou, S. Psych, M. R. C. (2005). The neurodevelopmental model of schizophrenia: update 2005. *Mol Psychiatry, 10,* 434-449.

Rice D. C. (1990). Lead-induced behavioral impairment on a spatial discrimination reversal task in monkeys exposed during different periods of development. *Toxicol Appl Pharmacol, 106*, 327-333.

Rice, D. C. & Gilbert, S. G. (1990). Effects of developmental exposure to methyl mercury on spatial and temporal visual function in monkeys. *Toxicol Appl Pharmacol, 102*, 151-163.

Rice, D. C. (1984). Behavioral deficit (delayed matching to sample) in monkeys exposed from birth to low levels of lead. *Toxicol Appl Pharmacol, 75,* 337-345.

Romero, E. Guaza, C. Castellano, B. Borrell, J. (2010). Ontogeny of sensorimotor gating and immune impairment induced by prenatal immune challenge in rats: implications for the etiopathology of schizophrenia. *Mol Psychiatry, 15,* 372-83.

Rosen, G. D. Bai, J. Wang, Y. Fiondella, C. G. Threlkeld, S. W. LoTurco, J. J. Galaburda, A. M. (2007). Disruption of neuronal migration by RNAi of Dyx1c1 results in neocortical and hippocampal malformations. *Cerebral Cortex, 17,* 2562-2572.

Roy, A. Bellinger, D. Hu, H. Schwartz, J. Ettinger, A. S. Wright, R. O. Bouchard, M. Palaniappan, K. Balakrishnan, K. (2009). Lead exposure and behavior among young children in Chennai, India. *Environ Health perspect, 117,* 1607-1611.

Sakata-Haga, H. Sawada, K. Hisano, S. Fukui, Y. (2001). Abnormalities of cerebellar foliation in rats prenatally exposed to ethanol. *Acta Neuropathol, 102*, 36-40.

Sandhir, R. Julka, D. Gill, K. D. (1994). Lipoperoxidative damage on lead exposure in rat brain and its implications on membrane bound enzymes. *Pharmacol Toxicol, 74*, 66-71.

Sano, S. Itami, S. Takeda, K. Tarutani, M. Yamaguchi, Y. Miura, H. Yoshikawa, K. Akira, S. Takeda, J. (1999). Keratinocyte-specific ablation of Stat3 exhibits impaired skin remodeling, but does not affect skin morphogenesis. *EMBO J, 18*, 4657-4668.

Sass, J. B. Haselow, D. T. Silbergeld, E. K. (2001). Methylmercury-induced decrement in neuronal migration may involve cytokine-dependent mechanisms: A novel method to assess neuronal movement in vitro. *Toxicol Sci, 63*, 74-81.

Segarra, J. Balenci, L. Drenth, T. Maina, F. Lamballe, F. (2006). Combined signaling through ERK, PI3K/AKT, and RAC1/p38 is required for met-triggered cortical neuron migration. *J Biol Chem, 281,* 4771-4778.

Serajee, F. J. Zhong, H. Mahbubul-Hug, A. H. (2006). Association of reelin gene polymorphisms with autism. *Genomics, 87,* 75-83.

Sham, P .C. MacLean, C. J. Kendler, K. S. (1993). Risk of schizophrenia and age difference with older siblings. Evidence for a maternal viral infection hypothesis? *Br J Psychiatry, 163,* 627-633.

Shi, L. Smith, S. E. Malkova, N. Tse, D. Su, Y. Patterson, P. H. (2009). Activation of the maternal immune system alters cerebellar development in the offspring. *Brain Behav Immun, 23,* 116-123.

Shinmura, Y. Kosugi, I. Aiba-Masago, S. Baba, S. Yong, L. R. Tsutsui, Y. (1997). Disordered migration and loss of viral-infected neuronal cells in developing mouse brains infected with murine cytomegalovirus. *Acta Neuropathol, 93,* 551-557.

Spalice, A. Parisi, P. Nicita, F. Pizzardi, G. Del Balzo, F. Iannetti, P. (2009). Neuronal migration disorders: clinical, neuroradiologic and genetics aspects. *Acta Pediatrica, 98,* 421-433.

Stehr-Green, P. Tull, P. Stellfeld, M. Mortenson, P. B. Simpson, D. (2003). Autism and thimerosal-containing vaccines: lack of consistent evidence for an association. *Am J Prev Med, 25,* 101-106.

Stein, J. Schettler, T. Wallinga, D. Valenti, M. (2002). In Harm's way: Toxic threats to child development. *Dev Behav. Pediatr, 23,* S13-S21.

Stipursky, J. & Gomes, F. C. (2007). TGF-beta1/SMAD signaling induces astrocyte fate commitment in vitro: implications for radial glia development. *Glia, 55*, 1023-1033.

Stumm, R. & Höllt, V. (2007) CXC chemokine receptor 4 regulates neuronal migration and axonal pathfinding in the developing nervous system: implications for neuronal regeneration in the adult brain. *J Mol Endocrinol, 38*, 377-382.

Sun, W. Funakoshi, H. Nakamura, T. (2002). Localization and functional role of hepatocyte growth factor (HGF) and its receptor c-met in the rat developing cerebral cortex. *Brain Res Mol Brain Res, 103,* 36-48.

Szabó, G. Katarova, Z. Körtvély, E. Greenspan, R. J. Urbán, Z. (1996). Structure and the promoter region of the mouse gene encoding the 67-kD form of glutamic acid decarboxylase. *DNA Cell Biol, 15,* 1081-1091.

Szelényi, J. (2001). Cytokines and the central nervous system. *Brain Res Bull, 54*, 329-338.

Taipale, M. Kaminen, N. Nopola-Hemmi, J. Haltia, T. Myllyluoma, B. Lyytinene, H. Muller, K. Kaaranen, M. Lindsberg, P. J. Hannula-Jouppi, K. Kere, J. (2003). A candidate gene for developmental dyslexia encodes a nuclear tetratricopeptide repeat domain protein dynamically regulated in brain. *Proc Natl Acad Sci USA, 100,* 11553-11558.

Takei, N. Mortensen, P. B. Klaening, U. Murray, R. M. Sham, P. C. O'Callaghan, E. Munk-JØrgensen, P. (1996). Relationship between in utero exposure to influenza epidemics and risk of schizophrenia in Denmark. *Biol Psychiatry, 40,* 817-824.

Takeuchi, T. (1977). Pathology of fetal Minamata disease: The effects of methylmercury on human interuterine life. *Pediatrician, 6*, 69-87.

Tanoue, Y. Oda, S. Asano, F. Kawashima, K. (1988). Epidemiology of infantile autism in southern Ibaraki, Japan: diferences in prevalence in birth cohorts. *J Autism Dev Disor, 18,* 155-156.

Tapia-Páez, I. Tammimies, K. Massinen, S. Roy, A. L. Kere, J. (2008). The complex of TFII-I, PARP1 and SFPQ proteins regulate the DYX1C1 gene implicated in neuronal migration and dyslexia. *FASEB J, 22,* 3001-3009.

Tasken, K. & Aandahl, E. M. (2004). Localized effects of cAMP mediated by distinct routes of protein kinase A. *Physiol Rev, 84*, 137-167.

Thanseem, I. Nakamura, K. Miyachi, T. Toyota, T. Yamada, S. Tsujii, M. Tsuchiya, K. J. Anitha, A. Iwayama, Y. Yamada, K. Hattori, E. Matsuzaki, H. Matsumoto, K. Iwata, Y. Suzuki, K. Suda, S. Kawai, M. Sugihara, G. Takebayashi, K. Takei, N. Ichikawa, H. Sugiyama, T. Yoshikawa, T. Mori, N. (2010). Further evidence for the role of MET in autism susceptibility. *Neurosci Res, 68,* 137-141.

Torrey, E. F. Barci, B. M. Webster, M. J. Bartko, J. J. Meador-Woodruff, J. H. Knable, M. B. (2005). Neurochemical markers for schizophrenia, bipolar disorder, and major depression in postmortem brains. *Biol Psychiatry, 57,* 252-260.

Toscano, C. D. and Guilarte T. R. (2005). Lead neurotoxicity: from exposure to molecular effects. *Brain Res Brain Res Rev, 49,* 529-554.

Tsutsui, Y. (2009). Effects of cytomegalovirus infection on embryogenesis and brain development. *Congenital Anomalies, 49*, 47-55.

Valiente, M. & Marín, O. (2010). Neuronal migration mechanisms in development and disease. *Curr Opin Neurobiol, 1,* 68-78.

Vargas, D.L. Nascimbene, C. Krishnana, C. Zimmerman, A.W. Pardo, A. (2005). Neuroglial activation and neuroinflammation in the brain of patients with autism. *Ann Neurol, 57,* 67-81.

Walker, L. M. Katzir, T. Liu, T. Ly, J. Corriveaux, K. Brazillai, M. Chu, F. O'Connor, M. G. Hackney, D. B. Chang, B. S. (2009). Gray matter volumes and cognitive ability in the epileptogenic brain malformation of periventricular nodular heterotopia. *Epilepsy Behav. 15,* 456-460.

Walter, H. J. & Messing, R. O. (1999). Regulation of neuronal voltage-gated calcium channels by ethanol. *Neurochem Int, 35,* 95-101.

Wang, H. L. Chen, X. T. Yang, B. Ma, F. L. Wang, S., Tang, M. L. Hao, M. G. Ruan, D. Y. (2008). Case-control study of blood lead levels and attention deficit hyperactivity disorder in Chinese children. *Enrion Health Perspect, 116,* 1401-1406.

Wegiel, J. Kuchna, I. Nowicki, K. Imaki, H. Wegiel, J. Marchi, E. Yong-Ma, S. Chauhan, A. Cauhan, V. Bobrowicz, T. W. de Leon, M. Saint Louis, L. A. Cohen, I. L. London, E. Brown, W. T. Wisniewski, T. (2010). The neuropathology of autism: defects of neurogenesis and neuronal migration, and dysplastic changes. *Acta Neuropathol, 119,* 755-770.

West, J. R. Chen, W. J. Pantazis, N. J. (1994). Fetal alcohol syndrome: the vulnerability of the developing brain and possible mechanisms of damage. *Metab Brain Dis, 9*, 291-322.

WHO/IPCS. 2004. Safety Evaluation of Certain Food Additives and Contaminants, Sixty-first meeting of the Joint FAO/WHO Expert Committee on Food Additives, WHO Food Additives Series, No. 52. Geneva: World Health Organization, International Programme on Chemical Safety.

Wilson, M. A. Johnston, M. V. Goldstein, G. W. Blue, M. E. (2000). Neonatal lead exposure impairs development of rodent barrel field cortex. *Proc Natl Acad Sci USA, 97*, 5540-5545.

Wisniewski, K. Dambska, M. Sher, J. H. Qazi, Q. (1983). A clinical neuropathological study of the fetal alcohol syndrome. *Neuropediatrics, 14,* 197-201.

Yamashita, Y. Fujimoto, C. Nakajima, E. Isagai, T. Matsuishi, T. (2003). Possible association between congenital cytomegalovirus infection and autistic disorder. *J Autism Dev Disord, 33,* 455-459.

Yan, E. Castillo-Meléndez, M. Nicholls, T. Hirst, J. Walker, D. (2004) Cerebrovascular responses in the fetal sheep brain to low-dose endotoxin. *Pediatr Res, 55,* 855-63.

Yee, S. Choi, B.H. (1996). Oxidative stress in neurotoxic effects of methylmercury poisoning. *Neurotoxicol, 17*, 17-26.

Zhao, S. & Frotscher, M. (2010). Go or stop? Divergent roles of Reelin in radial neuronal migration. *Neuroscientist, 16,* 421-434.

Zhou, K. Dempfle, A. Arcos-Burgos, M. Bakker, S. C., Banaschewski, T. Biederman, J. Buitelaar, J. Castellanos, F. X. Doyle, A. Ebstein, R. P. Ekholm, J. Forabosco, P. Franke, B. Freitag, C. Friedel, S. Gill, M. Hebebrand, J. Hinney, A. Jacob, C. Lesch, K.P. Loo, S. Lopera, F. McCracken, J.T. McGough, J.J. Meyer, J. Mick, E. Miranda, A. Muenke, M. Mulas, F. Nelson, S. F. Nguyen, T.T. Oades, R. D. Ogdie, M. N. Palacio, J. D. Pineda, D. Reif, A. Renner, T. J. Roeyers, H. Romanos, M. Rothenberger, A. Schafer, H. Sergeant, J. Sinke, R. J. Smalley, S. L. Sonuga-Barke, E. Steinhausen, H. C. Meulen, E. Walitza, S. Warnke, A. Lewis, C. M. Faraone, S. V. Asherson, P. (2008). Meta-analysis of genome-wide linkage scans of attention deficit hyperactivity disorder. *Am J Med Genet B Neuropsychiatr Genet, 147B,* 1392-1398.

In: Neuronal Migration
Editors: Andrew Girard and Luca Moreau
ISBN: 978-1-62100-059-4

Chapter II

Candidate Dyslexia Susceptibility Genes and Disorders of Neuronal Migration: Behavioral Effects of Cortical Dysgenesis in a Rodent Model

C. E. Szalkowski, R. H. Fitch
Department of Psychology, Behavioral Neuroscience Division;
University of Connecticut, Storrs, Connecticut, US

Abstract

Subtle alterations in the development of the neocortex are associated with cognitive deficits in humans and other mammals. A large number of genes modulate the development of the neocortex, and research into the behavioral phenotype associated with specific gene manipulations is advancing rapidly. Recent findings include evidence that anomalies in several human genes, (*DYX1C1, DCDC2, KIAA0319, MRPL19, C2ORF3, ROBO1, KIAA0319L*), may be associated with an increased incidence of dyslexia. Concurrent research has shown that the rat homologs for a subset of these genes (*DYX1C1, DCDC2, KIAA0319, ROBO1*) modulate critical parameters of early cortical development, including neuronal migration. Recent studies have utilized animal models of genetic disruption, using in utero RNA interference to knockdown expression of *Kiaa0319* and *Dyx1c1* in developing rat pups. Analyses have shown that interference with these genes leads to aberrant neuronal migration in the neocortex, resulting in the formation of ectopias and heterotopias in white matter. Interestingly, these malformations are similar to those observed in histological studies of brain tissue obtained *post mortem* from dyslexic individuals. Additionally, these genetic disruptions to neuronal migration have been associated with deficits in rapid auditory processing, spatial learning, and working memory in animal models. Parallel deficits in core processes critical to language and learning have been reported in dyslexic populations. Specifically, whereas dyslexia is diagnostically defined overall as a reading impairment in the absence of any generalized cognitive or neurologic deficit, this disorder likely reflects complex effects of multiple genes on multiple neural systems. As such, it is not surprising that more fine-grained behavioral sub-phenotypes have been identified within the dyslexic population (also called "intermediate phenotypes" in human genetic studies; see Marino *et al.*, 2007), and these may ultimately prove easier to link to the effects of specific dyslexia-risk genes. Reported sub-phenotypes include specific deficits in short term or working memory (verbal and also non-verbal), visual attention, and rapid auditory processing. It must be noted that the mapping of these intermediate phenotypes onto "clinical sub-types" of dyslexia (e.g., phonologic versus surface or orthographic dyslexia) has proved difficult—likely due to variations in definition, classification, and approach—but it seems likely that the clinical sub-types reported for dyslexia may reflect differential intermediate phenotypes and, in turn, the influence of different genes or gene combinations. One benefit of a focus on these intermediate phenotypes is that such behavioral measures are more amenable to modeling in animals (whereas reading *per se* is not). Thus, rodent models can be used in paradigms of working memory, visual attention and rapid auditory processing. Moreover, such rodent models can be employed to link specific intermediate phenotypes with individual dyslexia-risk genes. In the present review, behavioral results and histological data are discussed, and comparisons are made between the neuroanatomical and behavioral impacts of early interference with *Dyx1c1* and *Kiaa0319* in rodents. The recent findings are summarized and discussed in the context of improving our understanding of the genetic contributions to the behavioral expression of dyslexia in humans.

INTRODUCTION

Dyslexia and Neuronal Migration

Neocortical neuronal migration is the process by which newly divided neurons travel from the ventricular zone (VZ) where they are formed to their final target layer in the cerebral cortex. This delicate and important stage of neurodevelopment requires the perfect synchrony of several complex processes such as cell adhesion, cell signaling, and a vast array of dynamic cytoskeletal modifications (for review, see Hatten, 1999). Disruption to any of these processes can have detrimental effects on neuronal migration and thus to the development of the brain, and several well-known disorders are directly related to aberrant migration of neocortical neurons (Roberts, 2007; Spencer-Smith *et al.,* 2009; Valiente and Marin, 2010; Walsh *et al.,* 2000; Wynshaw-Boris *et al.*, 2010). In general, disruption to neuronal migration leads to misplaced neurons within and outside of the cortex, which often results in aberrant patterns of connectivity and synaptic transmission—ultimately creating atypical physiological and behavioral effects. In many disorders, the effects of disrupted neuronal migration are severe—often presenting as significant mental retardation, motor impairment, and in some cases even death (see Valiente and Marin, 2010; Wynshaw-Boris *et al.*, 2010 for review). Such disorders are perhaps not surprisingly accompanied by large scale changes to the structure of the brain that can easily be detected by MRI or CT scans.

Research within the last several decades has revealed that small, focal disruptions of neuronal migration are also associated with significant (though generally more specific and subtle) behavioral deficits. For example, Galaburda and colleagues (1985) examined *post mortem* brain tissue samples from individuals diagnosed with developmental dyslexia and discovered numerous focal dysplasias indicating early disruptions in the process of neuronal migration. Indeed, each of the four brains displayed several incidences of neuronal "nests" formed by neocortical neurons that had accumulated in small clusters within Layer 1 of the cortex. Along with these ectopic collections of cells, Galaburda *et al.* observed small instances of aberrant folding of the cortical tissue. These microgyria were characterized by fused laminae (only appearing to be 4 layered instead of 6), and a lack of columnar organization across the cortical layers in the affected tissue. Interestingly, the malformations were generally concentrated in perisylvian regions in the left hemisphere. Additionally, the observed cortical changes were accompanied by cell size changes in the medial geniculate nucleus, which had fewer large cells and more small cells than the MGN in control brains (Galaburda *et al.,* 1994). Although the cortical anomalies were not present in the exact same locations in each of the four brains, and some patients had more instances of disruption than others, Galaburda *et al.* concluded that these lesions had some pathogenetic role in the subjects' dyslexia. Authors suggested that the early cortical lesions may have led to widespread cortical reorganization, and in turn to the formation of anomalous connections. This was impossible to confirm in the *post mortem* tissue samples, but concurrent evidence supporting this proposed cortical rewiring comes from the use of animal models, which will be discussed later in this review. This was the first documented evidence suggesting a relationship between developmental dyslexia and aberrant neuronal migration. Unlike the neuroanatomy characteristic of many other neuronal migration disorders, the collections of aberrantly placed neurons in these brains were too small to be detected by current methods of *in vivo* imaging. This is perhaps not surprising given that developmental dyslexia—unlike disorders such as lissencephaly or double cortex syndrome—is not characterized by severe global impairments in learning. Instead, the deficits that define dyslexia are significant, but subtle and domain-specific.

Dyslexia Phenotypes: Rapid Auditory Processing

Developmental dyslexia is a complex disorder that is characterized by a variable constellation of underlying core behavioral deficits. The hallmark behavioral impairment in dyslexia is a deficit in reading, in spite of adequate intelligence and educational opportunity (Shaywitz, 1998). However, the behavioral deficits of dyslexia also encompass lower-order reading-related processes. For example, many dyslexic and language impaired individuals exhibit deficits in rapid auditory processing, which is the ability to detect changes in rapidly presented verbal or non-verbal acoustic cues. Tallal and colleagues were the first to demonstrate that verbal and non-verbal acoustic discrimination abilities were significantly correlated with performance on a nonword reading task (Tallal, 1980). These findings suggested that lower-level sensory processing abilities were associated with higher order speech and reading-related abilities. Several additional studies have since come out demonstrating a similar relationship between auditory processing and reading abilities in language impaired populations (Boets *et al.*, 2011; Boscariol *et al.*, 2010; Cohen-Mimran and Sapir, 2007; Czamara *et al.,* 2010; Gaab *et al.*, 2007; King *et al.*, 2008; Vandermosten *et al.*, 2011). Moreover, longitudinal studies by Benasich and colleagues have demonstrated that in children with a family history of language disability, deficits in the discrimination of two distinct tones (when presented at a short stimulus) can be elicited prior to emergent speech and reading capabilities, and serve as significant predictors of a future diagnosis of language impairment (Benasich *et al.*, 2002; 2006; Choudhury *et al.*, 2007). Taken together, these findings provide compelling evidence for the existence of a relationship between lower level sensory processing abilities and higher order language-related processes such as reading in dyslexia.

Dyslexia Phenotypes: Memory Impairment

Deficits in memory have also been reported in dyslexia. For example, deficits in verbal short term and working memory have been detected in some populations of dyslexics, and appear to be related to abilities such as word list recall—which is thought to tap into the phonological loop of working memory (Baddeley and Hitch, 1974). Similarly, language impaired individuals have been found to have impairments in verbal working memory: that is, the ability to process and store higher order verbal information that is required for sentence comprehension (Archibald and Gathercole 2006; Gathercole *et al.*, 2006; Smith-Spark and Fisk, 2007).

Deficits in visuospatial processing and visual memory, both of which are important for specific reading tasks, have also been occasionally reported in dyslexia. Dyslexic individuals have demonstrated impairments in visual pattern sequence recall as measured by the Corsi block span test and the visual patterns test (Smith-Spark and Fisk, 2007). Thus, the behavioral impairments that characterize dyslexia do not appear to be limited to verbal or strictly reading-related domains. It is also important to note that these deficits (and additional deficits not mentioned here, see Habib, 2000 for a complete review) are not represented homogeneously across all dyslexic individuals. In fact, a great deal of variability exists in the types and severity of impairment present from one individual to another. As such, some recent research into the etiology of developmental dyslexia has focused on the underlying behavioral deficits, or "intermediate phenotypes," that, in variable combinations with each other, make up a holistic diagnosis of dyslexia (Grigorenko, 2009).

Animal Models of Dyslexia: Behavioral Tasks

The interpretation of dyslexia as a clinical diagnosis overlying a pattern of causal intermediate phenotypes has opened the door for the use of animal models to study some of these lower-order behavioral processes, as well as their underlying neural substrates. For example, rapid auditory processing (RAP) is one behavioral facet of dyslexia that is amenable to animal modeling. In one set of studies, an operant conditioning paradigm was used to assess auditory processing capabilities of rats. Briefly, the rats were trained to discriminate and respond to a specific temporal order of a pair of high and low frequency tones in a go/no go paradigm. The interstimulus interval (ISI)—or the amount of time between the two tones in the pair—was gradually reduced over the course of testing to allow examination of the animals' acoustic processing abilities as a function of temporal parameters of the task (Fitch *et al.*, 1994).

More recent work from our lab has utilized a modified prepulse inhibition paradigm to assess temporal acoustic discrimination thresholds in rats as a function of attenuation of their acoustic startle reflex (see Fitch *et al.*, 2008a). The task relies on the basic tenets of prepulse inhibition, in which a startling stimulus is preceded by a non-startling prepulse or cue that the animal is able to detect and process, causing the animal's resulting acoustic startle reflex (a gross motor reflex) to be reduced in amplitude. During testing, animals are presented with a series of acoustic programs that are progressively more complex and difficult to discriminate, beginning with a basic silent gap detection task and progressing to a two-tone Oddball discrimination task and a frequency modulated (FM) sweep discrimination task. In all tasks, during cued trials the prepulse or cue is presented at variable rates, from long (easy to discriminate) to short (difficult to discriminate) interstimulus intervals. In all RAP tasks, an animal's relative ability to discriminate a cue is calculated as an "attenuation score" (ATT), which is a ratio of the amplitude of the animal's acoustic startle reflex during cued trials to the amplitude during uncued trials. A ratio of less than 1 (or 100%) indicates discrimination of the cue.

Fundamental short term and working memory abilities similar to those reported in dyslexia can also be tested using animal models. For example, recent work from our lab utilized a delayed match-to-sample radial arm water maze to assess rodent working memory abilities (see Chrobak *et al.*, 2008). Briefly, the task requires the animals to remember the position of a goal platform located at the end of one arm in an eight arm radial water maze. Each subject is initially presented with the location of the goal during a forced choice sample, in which all arms are blocked off besides the start arm and the goal arm (both of which varied systematically across days of testing). Following a retention interval of 10 minutes, the animal is returned to the maze in a new start position, and all of the arms are open. The animal's ability to retrieve the stored goal location from memory is measured as a function of the number of incorrect arms entered before finding the goal arm.

Animal Models of Dyslexia: Focal Freeze Lesions

Galaburda *et al.'s* findings of disrupted neuronal migration in the brains of dyslexics, combined with the ability to model some of the core behavioral processes involved in dyslexia, led to the development of subsequent animal models of migrational disruption. For example, one of the first animal models of "dyslexia" used in our lab was created via mechanical disruption to neuronal migration in rat pups. This was achieved through application of a freezing probe to the animal's skull plate on postnatal day 1 (P1), which produced a necrotic injury to the subjacent cortical tissue, characterized by the formation of a microgyric fold over time. The induced microgyric lesion was strikingly similar to those observed in the brains of dyslexic individuals, and was characterized by fused laminae and disrupted columnar organization (Rosen *et al.*, 1992). Anatomical studies of the brains of microgyric animals revealed widespread changes related to the disruption of neuronal migration, including significant changes in synaptic connections between the microgyric cortex and other areas of the cortex and the thalamus (Rosen *et al.*, 2000). Additional evidence for widespread changes following induction of microgyria came from another study that demonstrated that microgyria induced in the whisker barrel fields via focal freezing in P1 rat pups lead to an increase in size in the contralateral barrel field (Rosen *et al.,* 2001). Moreover, analysis of thalamic cells in microgyric animals revealed a shift in cell size distribution in the MGN of male animals following induction of the lesion in somatosensory cortex (Peiffer *et al.,* 2002a; Rosen *et al.*, 2006). The MGN of male microgyric animals was observed to have more small cells and fewer large cells than their sham counterparts—a finding highly similar to observations from dyslexic human *post mortem* brains (Galaburda *et al.,* 1994).

Given the neuroanatomical similarities between the brain tissue from dyslexic patients and microgyric rats, questions arose as to the behavioral phenotype of these animals. Subsequently, behavioral assessments of microgyric rats revealed interesting parallels to the clinical deficits observed in dyslexic humans. For example, when tested on the previously discussed operant conditioning go/no-go acoustic discrimination task, male microgyric rats showed impaired performance on short interstimulus interval trials only, indicating a specific impairment in rapid auditory processing (Clark *et al.,* 2000; Fitch *et al.*, 1994). Similarly, when tested on the modified prepulse inhibition task, male microgyric rats consistently showed less attenuation of their acoustic startle response (and thus less cue discrimination) when tested with rapidly presented cues (Peiffer *et al.*, 2002b; 2004a-c; Threlkeld *et al.,* 2006). Further behavioral studies of microgyric animals assessed the working memory abilities of lesioned animals, since working memory impairments are detected in some populations of dyslexics and language impaired individuals (Jeffries and Everatt, 2004; Smith-Spark and Fisk, 2007). A study using the previously described eight arm radial water maze found that male microgyric rats consistently made more incorrect arm entries before locating the goal platform throughout testing, indicating impaired spatial working memory as compared to shams (Fitch *et al.,* 2008b).

Animal Models of Dyslexia: NZB and BXSB Mice

More naturalistic animal models of disrupted neuronal migration similar to that observed in developmental dyslexia were derived from populations of inbred mice—specifically, NZB and BXSB mice—that displayed spontaneously occurring neocortical ectopias. The use of these mice allowed for detailed investigation of the electrophysiological properties of cells within Layer 1 ectopias, similar to those observed in the brains from human dyslexics (Galaburda *et al.,* 1994). These focal disruptions to neuronal migration were found to contain both glutamatergic and GABAergic neurons (Gabel and LoTurco, 2001), and cells within ectopias were found to receive both excitatory and inhibitory inputs from surrounding normatopic cortex. Tracing studies also revealed that ectopic cells formed aberrant connections with other regions of the cortex and the thalamus (Sherman *et al.,* 1990; Jenner *et al.*, 2000). Given the observed patterns of connectivity associated with these malformations, it is perhaps not surprising that they were shown to be vulnerable to epileptiform firing patterns (Gabel and LoTurco, 2001). These studies formed a critical foundation for understanding the electrophysiological properties (and potential functional implications) of cortical malformations similar to those observed in human dyslexic patients.

The behavioral profile of these ectopic mice is similar to that of microgyric rats. Studies have consistently demonstrated working memory impairments in ectopic mice when compared to non-ectopic littermates (Boehm *et al.*, 1996; Hyde *et al.*, 2001; Waters *et al.*, 1997). Additionally, male ectopic mice were significantly impaired compared to shams when discriminating rapidly presented acoustic cues, indicating temporally-specific acoustic processing impairments (Clark *et al.*, 2000; Peiffer *et al.*, 2002b). Studies of mice with spontaneously occurring ectopias thus provided a further link between focal disruptions of neuronal migration similar to those observed in the brains of dyslexics, and parallel behavioral deficits, using an animal model.

THE GENETICS OF DYSLEXIA

Currently, animal models of dyslexia are being created through the use of genetic manipulations. This work has been made possible by recent advances in epidemiological studies of developmental dyslexia, which have detected several regions on the human genome that are significantly associated with the disorder. The initial discovery that dyslexia is a heritable trait was derived from studies of twins, with concordance rates from 50-100% for dyslexia in monozygotic twins, in contrast to dizygotic twins' concordance rates of only 29-54% (Willcutt *et al.*, 2010). With the knowledge of dyslexia's heritability, more recent epidemiological studies of dyslexia have utilized genetic linkage studies to detect chromosomal loci that are inherited along with the dyslexia phenotype in affected families (Cardon *et al.*, 1994; Fisher *et al.*, 1999; Grigorenko *et al.*, 1997; 2000; Nopola-Hemmi *et al.*, 2000; Smith *et al.*, 1983). Genetic linkage studies have been followed up by genetic association studies, which seek out changes or mutations within specific genes located in the chromosomal loci identified in the linkage studies. Genetic association studies have been carried out in populations of dyslexics in several countries, and from these studies a handful of genes have been identified as candidate dyslexia susceptibility genes (CDSGs) (Francks *et al.*, 2004; Hannula-Jouppi *et al.*, 2005; Konig *et al.*, 2011; Meng *et al.*, 2005; Poelmans *et al.*, 2009; Scerri *et al.*, 2010; Taipale *et al.*, 2003). The first identified CDSG is located on Chromosome 15. Specifically, a translocation on the long arm of Chromosome 15 interrupting the genetic sequence of a gene—*DYX1C1*—was found to associate with specific reading disability in a single family in Finland (Taipale *et al.*, 2003). Further support for associations between *DYX1C1* and dyslexia has been reported in other populations as well (Brkanac *et al.*, 2007; Dahdouh *et al.*, 2009; Marino *et* al., 2007; Wigg *et al.*, 2004). However, there have also been reports of failed replication attempts (Bellini *et al.*, 2005; Marino *et al.*, 2005; Scerri *et al.*, 2004). Two additional CDSGs were identified on the short arm of Chromosome 6. Specifically, association analyses using markers for loci identified on Chromosome 6 revealed consistent correlations between dyslexia and single nucleotide polymorphisms (SNPs) within the sequences of the genes *KIAA0319* and *DCDC2* (Cope *et al.*, 2005; Elbert *et al.*, 2011; Francks *et al.*, 2004; Harold *et al.*, 2006; Ludwig *et al.*, 2007; Meng *et al.*, 2005; Newbury *et al.*, 2010; Schumacher *et al.*, 2006; Wilcke *et al.*, 2009). Additional CDSGs have recently been identified, but replications of these findings and further characterizations of the identified genes are limited (Hannula-Jouppi *et al.*, 2005; Konig *et al.*, 2011; Matsson *et al.*, 2011; Poelmans *et al.*, 2008; Scerri *et al.*, 2010). Thus, for the purposes of this review we will focus on findings concerning *DYX1C1* and *KIAA0319*, two well-studied candidate dyslexia risk genes.

Various behavioral assessments have been utilized in the genetic association studies of dyslexic populations, and some discernible patterns between impairments in specific behavioral domains and involvement of specific genes have emerged. For example, there is some evidence suggesting that variation in *DYX1C1* contributes to impairments in reading-related processes such as phonological awareness and phonological decoding—both of which characterize an individual's ability to break words down into smaller acoustic units (phonemes) (Wigg *et al.,* 2004). Additionally, a strong body of evidence has recently emerged indicating that *DYX1C1* likely plays a significant role in short term memory in dyslexics, as measured by tests of verbal short term memory such as nonword repetition and single letter backward span, (both of which require subjects to maintain acoustic information in short term memory during task performance; Dahdouh *et al.,* 2009; Marino *et al.,* 2007; Newbury *et al.,* 2010; Wigg *et al.,* 2004). A recent study in an Australian population also demonstrated that a mutation in *DYX1C1* was significantly associated with memory abilities in a non-affected group of individuals, thus indicating that the gene may be involved in core memory processes in populations of normal readers as well (Bates *et al.*, 2010). Taken together, these findings suggest a role for *DYX1C1* in the memory domain, which may account for its main contribution to the development of dyslexia.

Genotype-phenotype relationships between behavior and *KIAA0319* have also been recently reported. For example, various genetic association studies found significant correlations between variants of *KIAA0319* and impairments in reading-related tasks such as single word reading, phonological decoding, phonemic awareness, and orthographic coding (Cope *et al.,* 2005; Francks *et al.*, 2004; Harold *et al.,* 2006). Each of these tasks is phonologically based, and is dependent upon the subject's perception and understanding of letter sounds. Interestingly, there is also evidence that *KIAA0319* may play a role in normal variation in reading abilities in the general population as well (Luciano *et al.,* 2007; Paracchini *et al.,* 2008). In contrast to *DYX1C1,* there is no evidence for or against a role of *KIAA0319* in short term memory to date. Taken together, these data provide compelling evidence for a specific role of *KIAA0319* in phonological reading-related processes.

Animal Models of Dyslexia: RNA Interference

Following the advent of these advances in our genetic understanding of dyslexia, genetically altered animal models have been created to better characterize the rodent homologs of these genes and their protein products. Specifically, rats have been used to create models of a genetic knockdown of the rodent homologs of both *DYX1C1* and *KIAA0319.* In contrast to a genetic knockout, a genetic knockdown results in reduced amounts of a given gene's protein product, rather than a total loss. Moreover, some genetic knockdown methods allow researchers to assess the effects of genetic alterations in specific bodily systems—such as brain regions—as opposed to the system-wide effects attained with a complete genetic knockout. Recently, several researchers have used RNA interference (RNAi) to create targeted genetic knockdowns of the rodent homologs of *DYX1C1* and *KIAA0319* in populations of neocortical pyramidal progenitor neurons in the ventricular zone during fetal development. Briefly, the RNA interference method involves the creation of a short hairpin RNA (shRNA) plasmid that corresponds to a specific genetic sequence of interest in the rat genome (see LoTurco *et al.,* 2009 for review). When used to study genes in the brain, the shRNA vector is delivered to the population of migrating neurons in the brains of rats during embryonic development via injection into the lateral ventricles. Electroporation creates a temporary, favorable electrical environment that allows the shRNA vector to be taken into the nucleus of cells, where it is transcribed along with the cells' own genetic material. *In vivo*, the shRNA interferes with the translation of the protein that corresponds to the gene of interest. The RNAi method has provided researchers the unprecedented ability to characterize the brain-specific cellular and physiological functions of the Dyx1c1 and Kiaa0319 proteins (the rodent homologs of the human DYX1C1 and KIAA0319 proteins) in live models.

Brain expression profiles have shown that both *DYX1C1* and *KIAA0319* are expressed widely throughout the brain from early in development through adulthood in both rodents and humans. In rodents, *Dyx1c1* is expressed ubiquitously throughout the brain at modest levels, with the highest concentrations found in the neocortex, hippocampus, and choroid plexus (Rosen *et al.,* 2007). *KIAA0319* expression seems to be more specific to the brain than that of *DYX1C1* (which is expressed in several other tissues; see Rebhan *et al.*, 2011). Specifically, *KIAA0319* is expressed in the neocortex, ganglionic eminence, hippocampus, midbrain, and cerebellum of rats and humans (Paracchini *et al.,*2006). Animal models and cell culture lines have allowed researchers to investigate the cellular roles of the Dyx1c1 and Kiaa0319 proteins, both of which remain largely unknown. Recent investigations into the role of the Kiaa0319 protein have suggested that it may be involved in cell adhesion and cell signaling, both processes that play important roles during neuronal migration (Velayos-Baeza *et al.*, 2007; 2008; 2010). The potential role of the Kiaa0319 protein in cell adhesion is also supported by the observation that the physical neural-glia relationship—which is known to be mediated by cell adhesion proteins—is completely disrupted in neurons transfected with *Kiaa0319* shRNA. Following transfection, affected neurons exhibited orthogonal orientations to nearby radial glial fibers (Paracchini *et al.*, 2006). Work by Wang and colleagues suggested that the rodent homolog of DYX1C1 may be involved in regulation of cytoskeletal dynamics (Wang *et al.,* 2006). Early interference with *Dyx1c1* caused neurons to remain in the multipolar stage. Typically, migrating neurons transform from the multipolar stage to the bipolar stage. Thus it is possible that the Dyx1c1 protein is required for transformation out of the multipolar stage during migration. Other recent investigations into the cellular function of Dyx1c1 have also indicated a potential role for this protein in estrogen receptor metabolism (Massinen *et al.,* 2010). The potential functional implications of this role in terms of neurodevelopment are a topic of ongoing investigation.

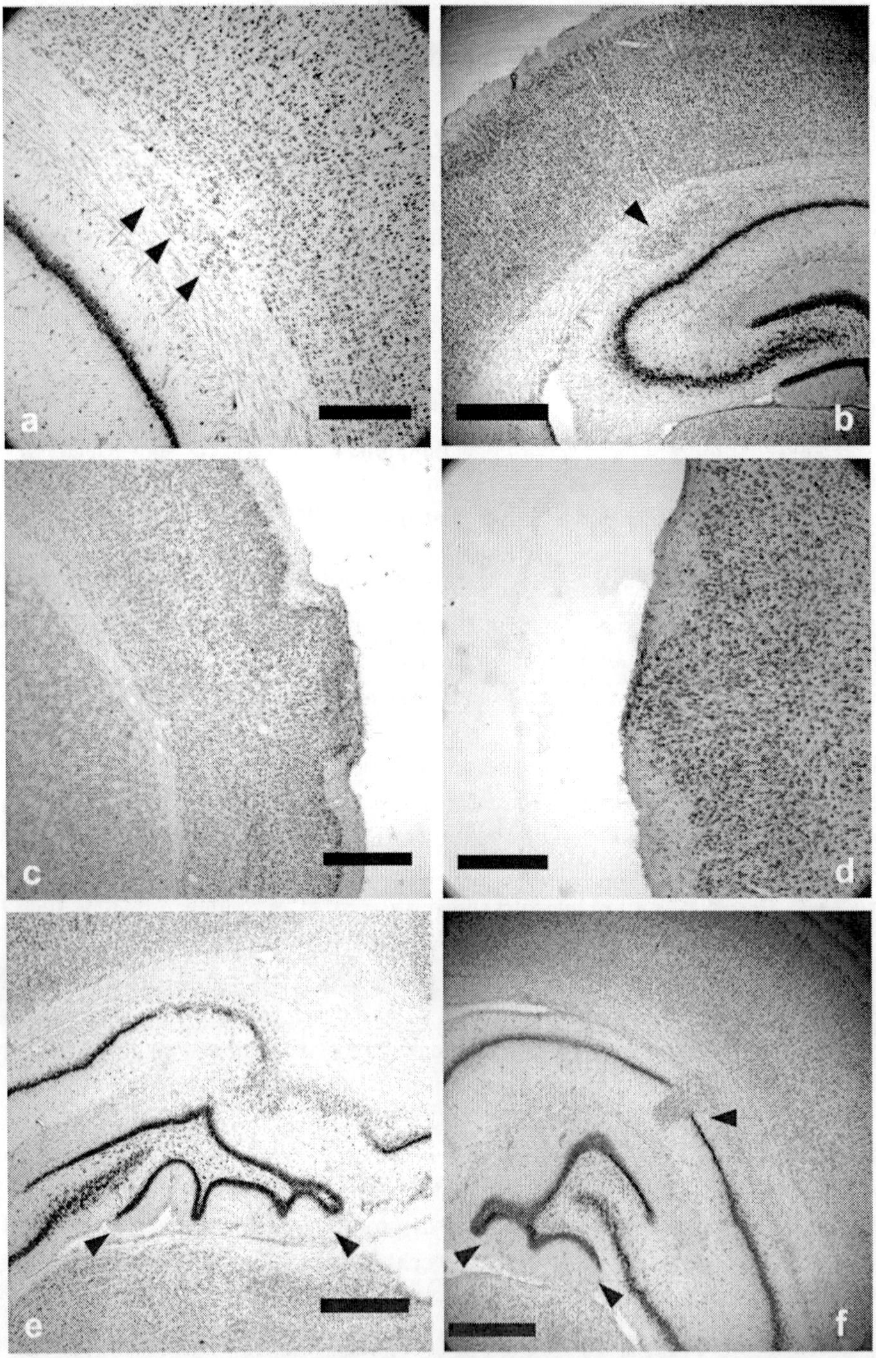

Figure 1. Neuroanatomical disruptions in rats following *in utero* transfection of *Dyx1c1* shRNA and *Kiaa0319* shRNA. Panels A, C, and E are representative images from coronal sections of adult rat brains following *in utero* transfection of *Dyx1c1* shRNA, while panels B, D, and F are representative images from rat brains following *in utero* transfection of *Kiaa0319* shRNA. All sections were stained for Nissl substance with cresyl violet. A., B. (Scale bars = 500 μm and 1 mm, respectively) Heterotopic collections of neurons in white matter (black arrows). C., D. (Scale bars = 1 mm) Layer I ectopia resulting from the injection wound during the transfection. E., F. (Scale bars = 1 mm) Hippocampal dysplasia, characterized by both heterotopic disruption of the CA1 region (black arrow) and abnormal folding of the dentate gyrus (black arrowheads).

Subsequent studies of these genes utilized the RNAi method to knockdown levels of each protein in the neocortex of developing rat pups, which led to the discovery that each protein plays a role in neuronal migration. For example, animals transfected embryonically with *Dyx1c1* shRNA showed significant delays in migration in a population of transfected Layer 2/3 pyramidal neurons, which were arrested in the white matter four days after the electroporation procedure (Wang *et al.,* 2006). In adult animals, a small subpopulation of transfected neurons were found to form heterotopic collections in the white matter and in Layer VI, while the majority of transfected neurons had migrated to supragranular layers—most of them traveling beyond their appropriate target layer (i.e., over migration; Rosen *et al.,* 2007). One quarter of the transfected animals also exhibited ectopic nests of neurons in Layer 1, similar to those previously observed in dyslexic brains (Galaburda *et al.,* 1985). Additionally, early interference with this gene created migrational anomalies within the hippocampus. Similarly, embryonic knockdown of *Kiaa0319* lead to an initial arrest of transfected neurons in the white matter (Paracchini *et al.,* 2007). Interestingly, the *Kiaa0319* shRNA transfection also caused a marked change in neuronal morphology, and the neuronal-radial glial cell interaction. That is, instead of being aligned parallel to one another with the migrating neuron's leading process attached to the radial glial fiber, the neuron attained a largely orthogonal orientation with respect to the radial glial cell (Paracchini *et al.,* 2007). The adult neuroanatomical phenotype following transfection of *Kiaa0319* shRNA is characterized by the presence of white matter heterotopia, abnormal laminar placement of individual transfected neurons, and migrational disruptions to the hippocampus (Peschansky *et al.,* 2009).

It is worth noting that some recent epidemiological studies have provided evidence that a subset of the variants of *KIAA0319* previously linked to dyslexia have base pair mutations that result in reduced (rather than eliminated) expression of the gene, which is similar to what is being modeled in the RNAi animals (Dennis *et al.,* 2009; Paracchini *et al.,* 2006). Although there is not currently an overwhelming body of evidence specifically linking reduced expression of CDSGs to dyslexia, the genetic knockdown rodent model has provided invaluable information on the developmental processes that these proteins are normally involved in, and the implications of disrupting these functions for subsequent morphology and behavior.

DYSLEXIA AND RNAI: RAPID AUDITORY PROCESSING

The epidemiological evidence of a role for *DYX1C1* and *KIAA0319* in dyslexia, combined with the evidence from animal models that these genes are involved in neurodevelopment and neuronal migration, led our lab to perform assessments on behaviors that parallel some of the intermediate phenotypes involved in dyslexia—namely, rapid auditory processing and working memory. [It is important to note that *post mortem* tissue analysis of all subjects following each behavioral study revealed consistent patterns of aberrant neocortical neuronal migration in shRNA treated animals, similar to those previously reported (Figure 1; see Peschansky *et al.*, 2009; Rosen *et al.*, 2007)]. Initially, we utilized the previously discussed modified pre-pulse inhibition paradigm to assess rapid and complex acoustic processing abilities of rats following embryonic interference with either *Dyx1c1* or *Kiaa0319* (see Figure 2).

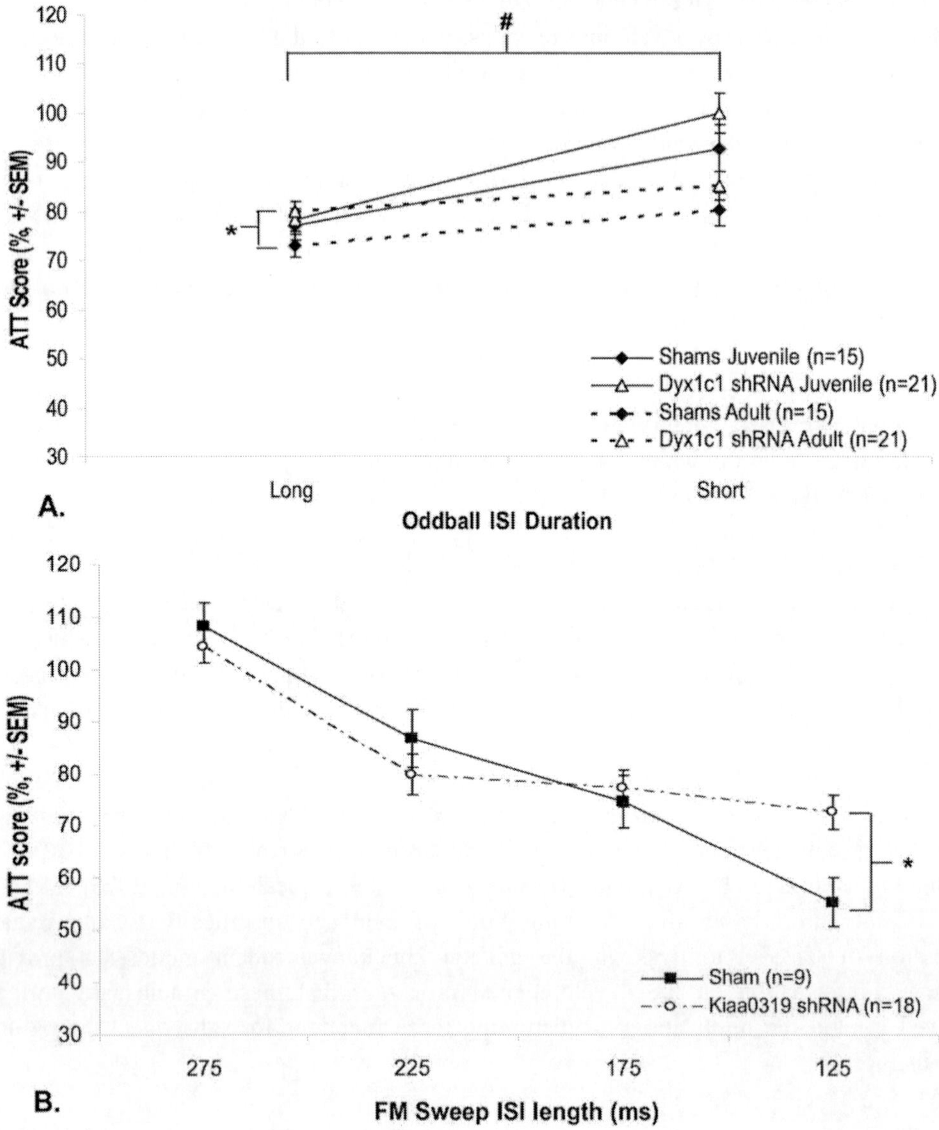

Figure 2. Auditory processing impairments in *Dyx1c1* and *Kiaa0319* shRNA treated animals. A. *Dyx1c1* shRNA animals: (adapted from Threlkeld *et al.*, 2007) On the Oddball test (a complex, two-tone frequency discrimination task), there was an overall near-significant Treatment effect (#, p = .051, one-tailed), suggesting that the *Dyx1c1* shRNA treated animals performed worse than shams overall on the task—at both long and short stimulus durations, as indicated by their higher attenuation scores. Additionally, there was a significant main effect of Treatment in adulthood on the long Oddball task [* indicates p < .05], with *Dyx1c1* shRNA animals performing worse than shams (Threlkeld *et al.*, 2007). B. *Kiaa0319* shRNA animals: (adapted from Szalkowski *et al.,* submitted) On FM Sweep (a complex frequency sweep discrimination task), there was a significant Treatment x ISI interaction (p < .05) and simple effects analysis revealed a significant Treatment effect at the shortest duration ISI, 125 ms, with *Kiaa0319* shRNA treated animals performing worse than shams (p < .01).

In the first study (assessing rapid auditory processing abilities of rats following transfection of *Dyx1c1* shRNA) we found that embryonic interference with *Dyx1c1* led to an auditory processing impairment. However, this impariment was not specific to rapid presentation of auditory stimuli. Rather, on an "Oddball task," (a complex, two-tone frequency discrimination task), an overall repeated measures ANOVA was found to show a near-significant Treatment effect ($p = .051$, one-tailed), suggesting that the *Dyx1c1* shRNA treated animals performed worse than shams overall on the task—at both long and short stimulus presentations. Additionally, there was a significant main effect of Treatment in adulthood on the long Oddball task [* indicates $p < .05$], with *Dyx1c1* shRNA animals performing worse than shams (Threlkeld *et al.,* 2007).

In a subsequent study testing the rapid auditory processing abilities in *Kiaa0319* shRNA treated rats, we found that on FM Sweep, a complex auditory processing task involving discrimination of a high-low frequency sweep, a repeated measures ANOVA revealed a significant Treatment x ISI interaction ($p < .05$). In this case, simple effects analysis revealed a significant Treatment effect only at the shortest duration ISI, 125 ms, with *Kiaa0319* shRNA treated animals performing worse than shams ($p < .01$). This indicated that adult *Kiaa0319* shRNA treated rats exhibited specific RAP deficits with rapid cue presentation on a complex frequency discrimination task (Szalkowski *et al.,* submitted).

Interestingly, in each case the transfected animals failed to show deficits on easier RAP tasks, such as silent gap detection. It was not until the animals were presented with more complex or rapid tasks (e.g., with temporal order discrimination, and/or stimuli at high rates (short durations) of presentation) that impairments in performance were evident. Moreover, it is interesting to speculate that there may be subtle differences in the RAP abilities of *Dyx1c1* and *Kiaa0319* transfected animals. Specifically, when presented with the complex FM sweep frequency discrimination task, *Kiaa0319* transfected animals showed significantly impaired detection only at the shortest interstimulus interval (i.e., the most rapid presentation of the acoustic cue (see Figure 2b)). This indicates that the auditory processing impairment induced by knockdown of *Kiaa0319* in these animals is temporally-specific. In contrast, when presented with a similar complex frequency discrimination RAP task (Oddball two-tone discrimination), the *Dyx1c1* transfected animals show impairment across long *and* short interstimulus intervals, suggesting that the deficit is not limited to processing of rapid stimuli, and may instead be a by-product of a disruption to core executive function or short term working memory (see Figure 2a).

DYSLEXIA AND RNAI: LEARNING AND SIMPLE SPATIAL MEMORY

Following each RAP study, animals were tested for basic spatial learning and memory abilities using a Morris water maze (Szalkowski *et al.,* submitted; Threlkeld *et al.,* 2007). The results of the Morris water maze testing are strikingly similar for the *Dyx1c1* and *Kiaa0319* studies, with significant impairment in task performance observed exclusively in those animals with specific disruptions to the hippocampus (see Figure 4). Interestingly, the presence of cortical malformations alone did not lead to significant spatial memory impairments in either the *Dyx1c1* or *Kiaa0319* shRNA animals, with cortically disrupted animals performing at a similar level to shams. In addition to providing further characterization of the functional effects of the genetically-induced hippocampal anomalies observed in these animals, the data derived from the Morris water maze testing serve as evidence that the shRNA treated animals are not "globally dysfunctional" as a result of their disrupted neurodevelopment, and thus their acoustic processing deficits are not simply the result of an all-encompassing impairment.

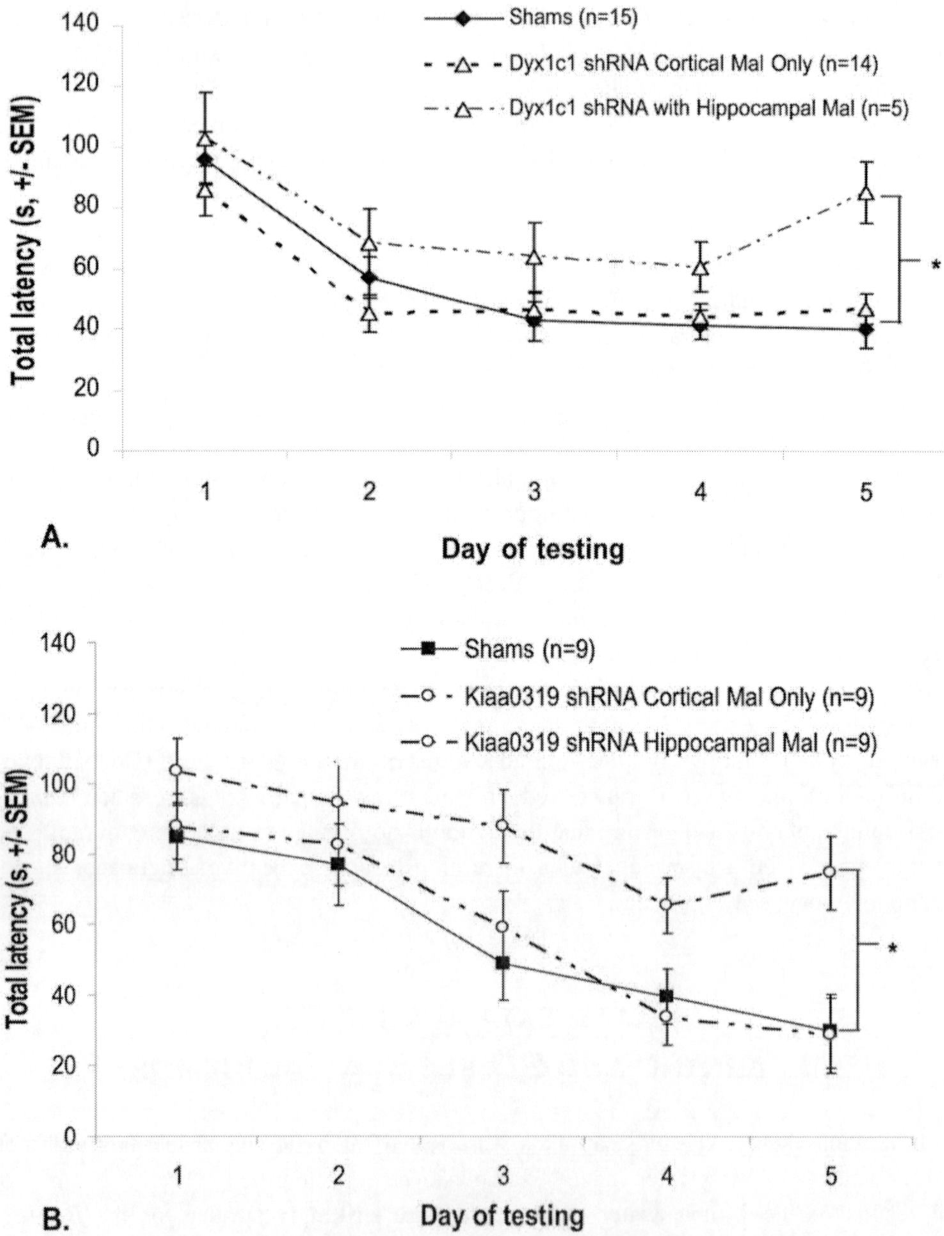

Figure 3. Simple spatial memory abilities in *Dyx1c1* and *Kiaa0319* shRNA treated rats. Simple spatial memory was assessed via Morris water maze testing. In each study, rats transfected with *Dyx1c1* shRNA or *Kiaa0319* shRNA with specific disruptions to the hippocampus took significantly longer to locate the platform during Morris water maze testing than both shams and shRNA treated animals without hippocampal malformations. (Adapted from Threlkeld *et al.*, 2007 and Szalkowski *et al.*, submitted, respectively; * indicates $p < .05$).

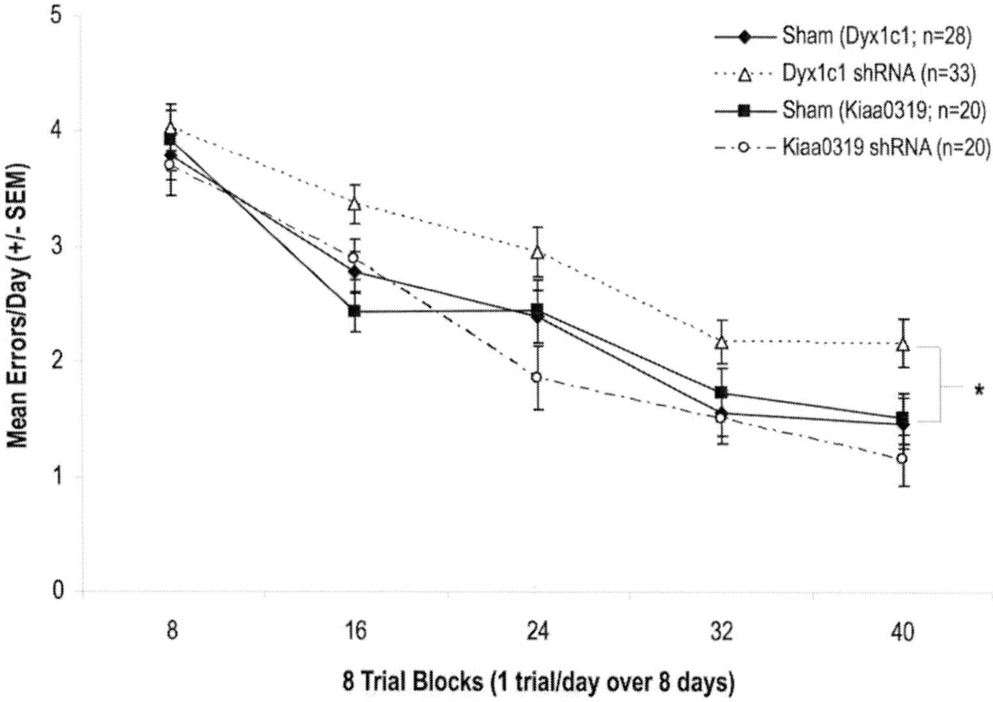

Figure 4. Working memory performance in *Dyx1c1* and *Kiaa0319* shRNA treated rats. Working memory abilities were assessed via delayed match-to-sample testing on an 8 arm radial water maze. *Dyx1c1* shRNA animals performed significantly worse than shams as indicated by an increased number of average incorrect arm entries (* indicates $p < .05$). Conversely, *Kiaa0319* shRNA treated animals did not exhibit any deficits on this task as compared to shams.

DYSLEXIA AND RNAI: WORKING MEMORY

Working memory was also assessed in groups of genetically disrupted animals using the previously discussed delayed match-to-sample 8 arm radial water maze task. Briefly, working memory performance is measured in this task as a function of the number of incorrect arm entries animals make as they attempt to navigate to a previously presented goal arm. In Study 1 assessing the effects of *Dyx1c1* knockdown on working memory, we found that across 12 weeks of testing *Dyx1c1* shRNA treated animals consistently made more incorrect arm entries during the test trial than shams, indicating a significant yet subtle impairment in working memory (Szalkowski *et al.*, 2011) (see Figure 4a). Interestingly, the results of Study 2, which assessed working memory in *Kiaa0319* shRNA treated animals, were strikingly different than those of Study 1. That is, throughout testing, the *Kiaa0319* shRNA treated animals performed the task equivalently to shams (Szalkowski *et al.*, submitted). Taken together, these data suggest that *Dyx1c1* may have a more salient role in working memory in rats as compared to *Kiaa0319*. This is especially interesting given two recent clinical findings that variants of the *DYX1C1* gene are specifically related to short term memory abilities in dyslexic humans (Dahdouh *et al.*, 2009; Marino *et al.,* 2007). This finding fits with the proposal that early interference with *Dyx1c1* may result in disruption to core executive function, which could result in impaired working memory and could cause the observed temporally-independent acoustic processing deficit, while *Kiaa0319* may have a more specific effect on temporal processing abilities. It is important to note that replications of these findings are necessary before firm conclusions can be drawn about these potential differences in the behavioral phenotypes associated with *Dyx1c1* and *Kiaa0319*.

CONCLUSION

The recent use of animal models to neuroanatomically model dyslexia has allowed researchers to discover links between disruption of neuronal migration and behavior. The even more recent advances in genetic studies of dyslexia (and the technology that allows for the creation of genetically altered animals) has brought us even closer to understanding the relationship between genes, neuroanatomy, and behavior as they relate to a complex cognitive disorder such as dyslexia in humans. The interpretation of dyslexia as a summation of variable intermediate phenotypes has been a key contributor to these recent advances. For example, strict examination of the intermediate phenotype of short term memory impairment, which is reported in some populations of dyslexics, has lead to a replicated association between *DYX1C1* and a specific, well-defined dyslexia-related behavior in different populations (Dahdouh *et al.,* 2009, Marino *et al.,* 2007). Moreover, relatively straightforward genotype-phenotype relationships such as the one between *DYX1C1* and short term memory are highly amenable to animal modeling. In closing, animal models provide an essential tool allowing researchers to draw connections between microscopic aspects of neuroanatomy and behavior that are not currently possible to attain through clinical studies alone. In the case of dyslexia, animal models also provide an avenue for defining relationships between risk genes and neuroanatomical and behavioral phenotypes, which could eventually lead to better definition of subtypes of dyslexia in humans. Further characterization of the candidate dyslexia genes may result in improved screening for the disorder, more accurate diagnoses, and ultimately the development of specifically tailored behavioral therapies and more effective interventions.

REFERENCES

Archibald, L.M.D. and Gathercole, S.E. (2006) Short-term and working memory in specific language impairment. *Int J Lang Comm Dis* 41, 675-693.

Baddeley, A.D. And Hitch, G. (1974) Working memory. *Psychol Learn Motiv* 8, 47-90.

Bates, T.C., Lind, P.A., Luciano, M., Montgomery, G.W., Martin, N.G., Wright, M.J. (2009) Dyslexia and *DYX1C1:* deficits in reading and spelling associated with a missense mutation. *Molec Psychiatr*, 1-7.

Bellini, G., Bravaccio, C., Calamoneri, F., Cocuzza, M.D., Fiorillo, P., Gagliano, A., Mazzone, D., Miraglia del Guidice, E., Scuccimarra, G., Militerni, R., Pascotto, A. (2005) No evidence for association between dyslexia and *DYX1C1* functional variants in a group of children and adolescents from Southern Italy. *J Molec Neurosci* 27, 311-314.

Benasich, A.A., and Tallal, P. (2002) Infant discrimination of rapid auditory cue predicts later language impairment. *Behav Brain Res* 136, 31-49.

Benasich, A.A., Choudhury, N., Friedman, J.T., Realpe-Bonilla, T., Chojnowski, C., Gou, Z. (2006) The infant as a prelinguistic model for language learning impairments: Predicting from event-related potentials to behavior. *Neuropsychologia* 44, 396-411.

Boehm, G.W., Sherman, G.F., Rosen, G.D., Galaburda, A.M., Denenberg, V.H. (1996) Neocortical ectopias in BXSB mice: Effects upon referens and working memory systems. *Cereb Cortex* 6, 696-700.

Boets, B., Vandermosten, M., Poelmans, H., Luts, H., Wouters, J., Ghesquiere, P. (2011) Preschool impairments in auditory processing and speech perception uniquely predict future reading problems. *Res Dev Disabil* 32, 560-570.

Boscariol, M., Guimaraes, C.A., Hage, S.R., Cendes, F., Guerreiro, M.M. (2010) Temporal auditory processing: correlation with developmental dyslexia and cortical malformation. *Pro Fono* 22, 537-542.

Brkanac, Z., Chapman, N.H., Matsushita, M.M., Chun, L., Nielsen, K., Cochrane, E., Berninger, V.W., Wijsman, E.M., Raskind, W.H. (2007) Evaluation of candidate genes for *DYX1* and *DYX2* in families with dyslexia. *Am J Med Genet B Neuropsychiatr Genet* 144B, 556-560.

Cardon, L.R., Smith, S.D., Fulker, D.W., Kimberling, W.J., Pennington, B.F., DeFries, J.C. (1994) Quantitative trait locus for reading disability on chromosome 6. *Science* 266, 276-279.

Choudhury, N., Leppanen, P.H.T., Leevers, H.J., Benasich, A.A. (2007) Infant information processing and family history of specific language impairment: converging evidence for RAP deficits from two paradigms. *Developmental Sci* 10, 213-236.

Chrobak, J.J., Hinman, J.R., Sabolek. (2008) Revealing past memories: Proactive interference and ketamine-induced memory deficits. *J Neurosci* 28, 4512-4520.

Clark, M.G., Sherman, G.F., Bimonte, H.A., Fitch, R.H. (2000) Perceptual auditory gap detection deficits in male BXSB mice with cerebrocortical ectopias. *NeuroReport* 11, 693-696.

Cohen-Mimran, R. and Sapir, S. (2007) Auditory temporal processing deficits in children with reading disabilities. *Dyslexia* 13, 175-192.

Cope, N., Harold, D., Hill, G., Moskvina, V., Stevenson, J., Holmans, P., Owen, M.J., O'Donovan, M.C., Williams, J. (2005) Strong evidence that *KIAA0319* on Chromosome 6p is a susceptibility gene for developmental dyslexia. *Am J Hum Genet* 76, 581-591.

Czamara, D., Bruder, J., Becker, J., Bartling, J., Hoffmann, P., Ludwig, K.U., Muller-Myhsok, B., Schulte-Korne, G. (2011) Association of a rare variant with mismatch negativity in a region between KIAA0319 and DCDC2 in dyslexia. *Behav Genet* 41,110-119.

Dahdouh, F., Anthoni, H., Tapia-Paez, I., Peyrard-Janvid, M., Schulte-Korne, G., Warnke, A., Remschmidt, H., Ziegler, A., Kere, J., Muller-Myhsok, B., Nothen, M.M., Schumacher, J., Zucchelli, M. (2009) Further evidence for *DYX1C1* as a susceptibility factor for dyslexia. *Psychiat Genet* 19, 59-63.

Dennis, M.Y., Paracchini, S., Scerri, T.S., Prokunina-Olsson, L., Knight, J.C., Wade-Martins, R., Coggill, P., Beck, S., Green, E.D., Monaco, A.P. (2009) A common variant associated with dyslexia reduces expression of the *KIAA0319* gene. *PloS Genetics* 5, e1000436.

Elbert, A., Lovett, M.W., Cate-Carter, T. , Pitch, A., Kerr, E.N., Barr, C.L. (2011) Genetic variation in the *KIAA0319* 5' region as a possible contributor to dyslexia. *Behav Genet* 41, 77-89.

Fisher, S.E., Marlow, A.J., Lamb, L., Maestrini, E., Williams, D.F., Richardson, A.J., Weeks, D.E., Stein, J.F., Monaco, A.P. (1999) A quantitative-trait locus on chromosome 6p influences different aspects of developmental dyslexia. *Am J Hum Genet* 64, 146-156.

Fitch, R.H., Tallal, P., Brown, C.P., Galaburda, A.M., Rosen, G.D. (1994) Induced microgyria and auditory temporal processing in rats: a model for language impairment? *Cereb Cortex* 4, 260-270.

Fitch, R.H., Threlkeld, S.W., McClure, M.M., Peiffer, A.M. (2008a) Use of a modified prepulse inhibition paradigm to assess complex auditory discrimination in rodents. *Brain Res Bull* 76, 1-7.

Fitch R.H., Breslawski, H., Rosen, G.D., Chrobak J.J. (2008b) Persistent spatial working memory deficits in rats with bilateral cortical microgyria. *Behav Brain Funct* 4, 45.

Francks, C., Paracchini, S., Smith, S.D., Richardson, A.J., Scerri, T.S., Cardon, L.R., Marlow, A.J., MacPhie, L., Walter, J., Pennington, B.F., Fisher, S.E., Olson, R.K., DeFries, J.C., Stein, J.F., Monaco, A.P. (2004) A 77-kilobase region of Chromosome 6p22.2 is associated with dyslexia in families from the United Kingdom and from the United States. *Am J Hum Genet* 75, 1046-1058.

Gaab, N., Gabrieli, J.D., Deutsch, G.K., Tallal, P., Temple, E. (2007) Neural correlates of rapid auditory processing are disrupted in children with developmental dyslexia and ameliorated with training: an fMRI study. *Restor Neurol Neurosci* 25, 295-310.

Gabel, L.A. And LoTurco, J.J. (2001) Electrophysiological and morphological characterization of neurons within neocortical ectopias. *J Neurophysiol* 85, 495-505.

Galaburda, A.M., Sherman, G.F., Rosen, G.D., Aboitiz, F., Geschwind, N. (1985) Developmental dyslexia: four consecutive patients with cortical anomalies. *Ann Neurol* 18, 222-233.

Galaburda, A.M., Menard, M.T., Rosen, G.D. (1994) Evidence for aberrant auditory anatomy in developmental dyslexia. *Proc Natl Acad Sci USA* 91, 8010-8013.

Gathercole, S.E., Alloway, T.P, Willis, C., Adams, A. (2006) Working memory in children with reading disabilities. *J Exper Child Psychol* 93, 265-281.

Grigorenko, E.L., Wood, F.B., Golovyan, L., Meyer, M.S., Romano, C., Pauls, D. (2000) Continuing the search for dyslexia genes on 6p. *Am J Med Genet B Neuropsychiatr Genet* 118B, 89-98.

Grigorenko, E.L., Wood, F.B., Meyer, M.S., Hart, L.A., Speed, W.C., Shuster, A., Pauls, D.L. (1997) Susceptibility loci for distinct components of developmental dyslexia on chromosomes 6 and 15. *Am J Hum Genet* 60, 27-39.

Grigorenko, E.L. (2009) At the height of fashion: what genetics can teach us about neurodevelopmental disabilities. *Curr Opin Neurol* 22, 126-130.

Habib, M. (2000) The neurological basis of developmental dyslexia. *Brain* 123, 2373-2399.

Hannula-Jouppi, K., Kaminen-Ahola, N., Taipale, M., Eklund, R., Nopola-Hemmi, J., Kaariainen, H., Kere, J. (2005) The axon guidance receptor gene *ROBO1* is a candidate gene for developmental dyslexia. *PLoS Genet* 1, e50.

Harold, D., Paracchini, S., Scerri, T., Dennis, M., Cope, N., Hill, G., Moskvina, V., Walter, J., Richardson, A.J., Owen, M.J., Stein, J.F., Green, E.D., O'Donovan, M.C., Williams, J., Monaco, A.P. (2006) Further evidence that the *KIAA0319* gene confers susceptibility to developmental dyslexia. *Molec Psychiatr* 11, 1085-1091.

Hatten, M.E. (1999) Central nervous system neuronal migration. *Annu Rev Neurosci*, 22, 511-539.

Hyde, L.A., Hoplight, B.J., Harding, S., Sherman, G.F., Mobraaten, L.E., Denenberg, V.H. (2001) Effects of ectopias and their cortical location on several measures of learning in BXSB mice. *Dev Psychobiol* 39, 286-300.

Jeffries, S. And Everatt, J. (2004) Working memory: Its role in dyslexia and other specific learning difficulties. *Dyslexia* 10, 196-214.

Jenner, A.R., Galaburda, A.M. Sherman, G.F. (2000) Connectivity of ectopic neurons in the molecular layer of the somatosensory cortex in autoimmune mice. *Cereb Cortex* 10, 1005-1013.

King, B., Wood, C., Faulkner, D. (2008) Sensitivity to visual and auditory stimuli in children with developmental dyslexia. *Dyslexia* 14, 116-141.

Konig, I.R., Schumacher, J., Hoffmann, P., Kleensang, A., Ludwig, K.U., Grimm, T., Neuhoff, N., Preis, M., Roeske, D., Warnke, A., Propping, P., Remschmidt, H., Nothen, M.M., Ziegler, A., Muller-Myhsok, B., Schulte-Korne, G. (2011) Mapping for dyslexia and related cognitive trait loci provides strong evidence for further risk genes on Chromosome 6p21. *Am J Med Genet B Neuropsychiatr Genet* 156, 36-43.

LoTurco, J.J., Manent, J.B., Sidiqi, F. (2009) New and improved tools for in utero electroporation studies of developing cerebral cortex. *Cereb Cortex* 19, i120-i125.

Ludwig, K.U., Roeske, D., Schumacher, J., Schulte-Korne, G., Konig, I.R., Warnke, A., Plume, E., Ziegler, A., Remschmidt, H., Muller-Myhsok, B., Nothen, M.M., Hoffmann, P. (2008) *J Neural Transm* 115, 1587-1589.

Marino, C., Giorda, R., Lorusso, M.L., Vanzin, L., Salandi, N., Nobile, M., Citterio, A., Beri, S., Crespi, V., Battaglia, M., Molteni, M. (2005) A family-based association study does not support *DYX1C1* on 15q21.3 as a candidate gene in developmental dyslexia. *Eur J Hum Genet* 13, 491-499.

Marino, C., Citterio, A., Giorda, R., Facoetti, A., Menozzi, G., Vanzin, L., Lorusso, M.L., Nobile, M., Molteni, M. (2007) Association of short-term memory with a variant within *DYX1C1* in developmental dyslexia. *Genes Brain Beh* 6, 640-646.

Massinen, S., Tammimies, K., Tapia-Paez, I., Matsson, H., Hokkanen, M.E., Soderberg, O., Landegren, U., Castren, E., Gustafsson, J.A., Treuter, E., Kere, J. (2009) Functional interaction of DYX1C1 with estrogen receptors suggests involvement of hormonal pathways in dyslexia. *Hum Molec Genet* 18, 2802-2812.

Matsson, H., Tammimies, K., Zucchelli, M., Anthoni, H., Onkamo, P., Nopola-Hemmi, J., Lyytinen, H., Leppanen, P.H.T., Neuhoff, N., Warnke, A., Schulte-Korne, G., Schumacher, J., Nothen, M.M., Kere, J., Peyrard-Janvid, M. (2011) SNP variations in the 7q33 region containing *DGKI* are associated with dyslexia in the Finnish and German populations. *Behav Genet* 41, 134-140.

Meng, H., Smith, S.D., Hager, K., Held, M., Liu, J., Olson, R.K., Pennington, B.F., DeFries, J.C., Gelernter, J., O'Reilly-Pol, T., Somlo, S., Skudlarski, P., Shaywitz, S.E., Shaywitz, B.A., Marchione, K., Wang, Y., Paramasivam, M., LoTurco, J.J., Page, G.P., Gruen, J.R. (2005) *Proc Natl Acad Sci USA* 102, 17053-17058.

Newbury, D.F., Paracchini, S., Scerri, T.S., Winchester, L., Addis, L., Richardson, A.J., Walter, J., Stein, J.F., Talcott, J.B., Monaco, A.P. (2011) Investigation of dyslexia and SLI risk variants in reading- and language-impaired subjects. *Behav Genet* 41, 90-104.

Nopola-Hemmi, J., Taipale, M., Haltia, T., Lehesjoki, A.E., Voutilainen, A., Kere, J. (2000) Two translocations of chromosome 15q associated with dyslexia. *J Med Genet* 37, 771-775.

Paracchini, S., Thomas, A., Castro, S., Lai, C., Paramasivam, M., Wang, Y., Keating, B.J., Taylor, J.M., Hacking, D.F., Scerri, T., Francks, C., Richardson, A.J., Wade-Martins, R., Stein, J.F., Knight, J.C., Copp, A.J., LoTurco, J.J., Monaco, A.P. (2006) The chromosome 6p22 haplotype associated with dyslexia reduces the expression of *KIAA0319*, a novel gene involved in neuronal migration. *Hum Mol Genet* 15, 1659-1666.

Paracchini, S., Scerri, T., Monaco, A.P. (2007) The genetic lexicon of dyslexia. *Annu Rev Geno Human Genet* 8, 57-79.

Paracchini, S., Steer, C.D., Buckingham, L., Morris, A.P., Ring, S., Scerri, T., Stein, J., Pembrey, M.E., Ragoussis, J., Golding, J., Monaco, A.P. (2008) Association of the *KIAA0319* dyslexia susceptibility gene with reading skills in the general population. *Am J Psychiatry* 165, 1576-1584.

Peiffer, A.M., Rosen, G.D., Fitch, R.H. (2002a) Rapid auditory processing and MGN morphology in microgyric rats reared in varied acoustic environments. *Dev Brain Res* 138, 187-193.

Peiffer, A.M., Rosen, G.D., Fitch, R.H. (2002b) Sex differences in rapid auditory processing deficits in ectopic BXSB/MpJ mice. *NeuroReport* 13, 2277-2280.

Peiffer, A.M., Friedman, J.T., Rosen, G.D., Fitch, R.H. (2004a) Impaired gap detection in juvenile microgyric rats. *Dev Brain Res* 152, 93-98.

Peiffer, A.M., Rosen, G.D., Fitch, R.H. (2004b) Sex differences in rapid auditory processing deficits in microgyric rats. *Dev Brain Res* 148, 53-57.

Peiffer, A.M., McClure, M.M., Threlkeld, S.W., Rosen, G.D., Fitch, R.H. (2004c) Severity of focal microgyria and associated rapid auditory processing deficits. *NeuroReport* 15, 1923-1926.

Peschansky, V.J., Burbridge, T.J., Volz, A.J., Fiondella, C.G., Wissner-Gross, Z., Galaburda, A.M., LoTurco, J.J., Rosen, G.D. (2010) *Cereb Cortex* 20, 884-897.

Poelmans, G., Engelen, J.J.M., Van Lent-Albrechts, J., Smeets, H.J., Schoenmakers, E., Franke, B., Buitelaar, J.K., Wuisman-Frerker, M., Erens, W., Steyaert, J., Schrander-Stumpel, C. (2009) Identification of novel dyslexia candidate genes through the analysis of a chromosomal deletion. *Am J Med Genet Part B* 150B, 140-147.

Rebhan, M., Chalifa-Caspi, V., Prilusky, J., Lancet, D.: GeneCards: encyclopedia for genes, proteins and diseases. Weizmann Institute of Science, Bioinformatics Unit and Genome Center (Rehovot, Israel), 1997. GeneCard for DYX1C1 23 May 2011. http://www.genecards.org/cgi-bin/carddisp.pl?gene=DYX1C1

Roberts, R.C. (2007) Disrupted in Schizophrenia (*DISC1*): Integrating clinical and basic findings. *Schizophrenia Bull* 33, 11-15.

Rosen, G.D., Press, D.M., Sherman, G.F., Galaburda, A.M. (1992) The development of induced cerebrocortical microgyria in the rat. *J Neuropathol Exp Neurol* 51, 601-611.

Rosen, G.D., Burstein, D., Galaburda, A.M. (2000) Changes in efferent and afferent connectivity in rats with induced cerebrocortical microgyria. *J Comp Neurol* 418, 423-440.

Rosen, G.D., Windzio, H., Galaburda, A.M. (2001) Unilateral induced neocortical malformation and the formation of ipsilateral and contralateral barrel fields. *Neuroscience* 103, 931-939.

Rosen, G.D., Mesples, B., Hendriks, M., Galaburda, A.M. (2006) Histometric changes and cell death in the thalamus after neonatal neocortical injury in the rat. *Neuroscience* 141, 875-888.

Rosen, G.D., Bai, J., Wang, Y., Fiondella, C.G., Threlkeld, S.W., LoTurco, J.J., Galaburda, A.M. (2007) Disruption of neuronal migration by RNAi of *Dyx1c1* results in neocortical and hippocampal malformations. *Cereb Cortex* 17, 2562-2572.

Scerri, T.S., Fisher, S.E. Francks, C., MacPhie, I.L., Paracchini, S., Richardson, A.J., Stein, J.F., Monaco, A.P. (2004) Putative functional alleles of *DYX1C1* are not associated with dyslexia susceptibility in a large sample of sibling paris from the UK. *J Med Genet* 41, 853-857.

Scerri, T.S., Paracchini, S., Morris, A., MacPhie, I.L., Talcott, J., Stein, J., Smith, S.D., Pennington, B.F., Olson, R.K., DeFries, J.C., Monaco, A.P. (2010) Identification of candidate genes for dyslexia susceptibility on Chromosome 18. *PLoS ONE* 5, e13712.

Schumacher, J., Anthoni, H., Dahdouh, F., Konig, I.R., Hillmer, A.M., Kluck, N., Manthey, M., Plume, E., Warnke, A., Remschmidt, H., Hulsmann, J., Cichon, S., Lindgren, C.M., Propping, P., Zucchelli, M., Ziegler, A., Peyrard-Janvid, M., Schulte-Korne, G., Nothen, M.M., Kere, J. (2006) Strong genetic evidence of *DCDC2* as a susceptibility gene for dyslexia. *Am J Hum Genet* 78, 52-62.

Shaywitz, S.E. (1998) Current concepts: Dyslexia. *New Engl J Med* 338, 307-312.

Sherman, G.F., Stone, J.S., Press, D.M., Rosen, G.D., Galaburda, A.M. (1990) Abnormal architecture and connections disclosed by neurofilament staining in the cerebral cortex of autoimmune mice. *Brain Res* 529, 202-207.

Smith, S.D., Kimberling, W.J., Pennington, B.F., Lubs, H.A. (1983) Specific reading disability: Identification of an inherited from through linkage analysis. *Science* 219, 1345-1347.

Smith-Spark, J.H. and Fisk, J.E. (2007) Working memory functioning in developmental dyslexia. *Memory* 15, 34-56.

Spencer-Smith, M., Leventer, R., Jacobs, R., de Luca, C., Anderson, V. (2009) Neuropsychological profile of children with subcortical band heterotopia. *Dev Med Child Neurol*, 51, 909-916.

Szalkowski, C.E., Hinman, J.R., Threlkeld, S.W., Wang, Y., LePack, A., Rosen, G.D., Chrobak, J.J., LoTurco, J.J., Fitch, R.H. (2011) Persistent spatial working memory deficits in rats following *in utero* RNAi of *Dyx1c1*. *Genes Brain Beh* **10**, 244-252.

Szalkowski, C.E., Fiondella, C.G., Galaburda, A.M., Rosen, G.D., LoTurco, J.J., Fitch, R.H. Neocortical disruption and behavioral impairments in rats following in utero RNAi of candidate dyslexia risk gene *Kiaa0319*, submitted.

Taipale, M., Kaminen, N., Nopola-Hemmi, J., Haltia, T., Myllyluoma, B., Lyytinen, H., Muller, K., Kaaranen, M., Lindsberg, P.J., Hannula-Jouppi, K., Kere, J. (2003) A candidate gene for developmental dyslexia encodes a nuclear tetratricopeptide repeat domain protein dynamically regulated in brain. *Proc Natl Acad Sci USA* 100, 11553-11558.

Tallal, P. (1980) Auditory temporal perception, phonics and reading disabilities in children. *Brain Lang,* 9, 182–198.

Threlkeld, S.W., McClure, M.M., Rosen, G.D., Fitch, R.H. (2006) Developmental timeframes for induction of microgyria and rapid auditory processing deficits in the rat. *Brain Res Bull* 1109, 22-31.

Threlkeld, S.W., McClure, M.M., Bai, J., Wang, Y., LoTurco, J.J., Rosen, G.D., Fitch, R.H. (2007) Developmental disruptions and behavioral impairments in rats following *in utero* RNAi of *Dyx1c1*. *Brain Res Bull* 71, 508-514.

Valiente, M. and Marin, O. (2010) Neuronal migration mechanisms in development and disease. *Curr Opin Neurobiol* 20, 68-78.

Vandermosten, M., Boets, B., Luts, H., Poelmans, H., Wouters, J., Ghesquiere, P. (2011) Impairments in speech and nonspeech sound categorization in children with dyslexia are driven by temporal processing difficulties. *Res Dev Disabil* 32, 593-603.

Velayos-Baeza, A., Toma, C., da Roza, S., Paracchini, S., Monaco, A.P. (2007) Alternative splicing in the dyslexia-associated gene *KIAA0319*. *Mamm Genome* **9**, 627-634.

Velayos-Baeza, A., Toma, C., Paracchini, S., Monaco, A.P. (2008) The dyslexia-associated gene *KIAA0319* encodes highly N- and O-glycosylated plasma membrane and secreted isoforms. *Hum Mol Genet* **17**, 859-871.

Velayos-Baeza, A., Levecque, C., Kobayashi, K., Holloway, Z.G., Monaco, A.P. (2010) The dyslexia-associated KIAA0319 protein undergoes proteolytic processing with {gamma}-secretaste-independent intramembrane cleavage. *J Biol Chem* **285**; 40148-40162.

Walsh, C.A. (2000) Genetics of neuronal migration in the cerebral cortex. *Ment Retard Dev D R*, 6, 34-40.

Wang, Y., Paramasivam, M., Thomas, A., Bai, J., Kaminen-Ahola, N., Kere, J., Voskuil, J., Rosen, G.D., Galaburda, A.M., LoTurco, J.J. (2006) Dyx1c1 functions in neuronal migration in developing neocortex. *Neuroscience* 143, 515-522.

Waters, N.S., Sherman, G.F., Galaburda, A.M., Denenberg, V.H. (1997) Effects of cortical ectopias on spatial delayed-matching-to-sample performance in BXSB mice. *Beh Brain Res* 84, 23-29.

Wigg, K.G., Couto, J.M., Feng, Y., Anderson, B., Cate-Carter, T.D., Macciardi, F., Tannock, R., Lovett, M.W., Humphries, T.W., Barr, C.L. (2004) Support for *EKN1* as the susceptibility locus for dyslexia on 15q21. *Molec Psychiatr* 9, 1111-1121.

Wilcke, A., Weissfuss, J., Kirsten, H., Wolfram, G., Boltze, J., Ahnert, P. (2009) The role of gene DCDC2 in German dyslexics. *Ann of Dyslexia* 59, 1-11.

Willcutt, E.G., Pennington, B.F., Duncan, L., Smith, S.D., Keenan, J.M., Wadsworth, S., DeFries, J.C., Olson, R.K. (2010) Understanding the complex etiologies of developmental disorders: Behavioral and molecular genetic approaches. *J Dev Behav Pediatr* 31, 533-544.

Wynshaw-Boris, A., Pramparo, T., Youn, Y.H., Hirotsune, S. (2010) Lissencephaly: Mechanistic insights from animal models and potential therapeutic strategies. *Sem Cell Dev Biol* 21, 823-830.

In: Neuronal Migration
Editors: Andrew Girard and Luca Moreau
ISBN: 978-1-62100-059-4

Chapter III

Electrically Guiding Nerve Growth and Neuronal Migration

***Lei Zhang*[1,2]*, Zhengping Yu*[2]*, Min Zhao*[1,*]**
[1]Institute for Regenerative Cures, Department of Dermatology and Department of Ophthalmology, University of California Davis, Sacramento, California, US
[2]Department of Occupational Health, Third Military Medical University, Chongqing, China

Abstract

Guided and stimulated nerve growth and neuronal migration is critical to repairing diseased or damaged nervous systems. Electric fields (EFs) provide a powerful signal to stimulate and guide nerve growth and neuronal migration. EFs may direct axon outgrowth, orientate cell polarity, and guide neuronal migration. These effects may be mediated through activation of important intracellular signaling pathways and orientation of important organelles. The potential use of EFs in the nervous system may provide new therapies for neurodevelopmental disorders and brain injuries. This chapter aims to review recent experimental data on EFs-guided nerve growth and neuronal migration, and possible molecular and cellular mechanisms.

1. Introduction

Directional migration of neuronal cells over various distances is critical to developing and regenerating nervous systems. Neural stem cells (NSCs) and neuroblasts reside in specific locations in the central nervous system (CNS). Newly born neurons are often far away from their sites of destiny, or sites of damage. Effective and directional neuronal

[*] Corresponding author: Min Zhao, Institute for Regenerative Cures, 2921 Stockton Blvd., University of California Davis, Sacramento, California 95817, USA. Tel: 1 916 703 9381; Fax: 1 916 703 9384; Email: minzhao @ucdavis.edu.

migration is therefore crucial to the development of the CNS, and effective neurogenesis and repair following neurodegeneration or injury to nervous system [1, 2].

Many guidance signals participate in the control of directional nerve growth and neuronal migration. These signals include the most well-studied ones such as chemical gradients, either soluble molecules or extracellular matrix molecules, which may be setup by diffusion or the flow of cerebral spinal fluid [3, 4].

A role for EFs in guiding neurite growth has been proposed and demonstrated in many types of nerve cells, in addition to the well-accepted guidance cues such as chemical gradients. It has been known for over a century that nerve growth, i.e. neurite growth can be affected by applied EFs. Some recent studies demonstrate that neuronal cell body migration can also be guided by small applied EFs. Guidance of migrating neurons shares many features with axon guidance, from the use of substrates to the specific cues regulating chemo-taxis/chemo-repulsion.

We will briefly review the recent experimental results focusing on electrically guided neuronal migration with particular reference to axon guidance.

2. Endogenous EFs in the Nervous Systems

Endogenous EFs may play an important role in neuronal development. Naturally occurring EFs have been implicated in guiding cell migration during embryonic development. Electric currents were detected in the embryos of *Xenopus* [5-7], axolotl [8, 9], and chick embryos [10-12]. In *Xenopus* embryos, endogenous EFs can reach up to 40 mV/mm [6]. Mapping of *Xenopus* embyo electric potentials revealed that the blastopore was the only site of outward current, and all other loci had inward currents which varied from 0.1 $\mu A/cm^2$ to several $\mu A/cm^2$. The neural tube of *Xenopus* embryos also maintains an electric potential of 18±1 mV, with the lumen being negative relative to the interstitial spaces. This electric potential varied between different developmental stages of the embryos [5]. If the embryo was damaged unintentionally, a large outward current associated with the wound site, was detected [7]. Similar results were obtained in chick embryos [10]. The region of outward current is a site of tissue involution, with the cells migrating in a direction opposite to the electric-field vector.

Disturbances to endogenous physiological EFs resulted in abnormal development of the nervous system [8]. The endogenous electric currents of embryos are composed by ion currents. Na^+ and Ca^{2+} fluxes have been shown to be the major contributors to the current [6, 7, 13]. Na^+ channel blockers, amiloride and benzamil, greatly reduced the current across the *Xenopus* embryo [7] and axolotl [8], which also resulted in failure of normal development, with cranial enlargement near or at the region of iontophoresis of the drugs. Chick embryos exhibit some developmental abnormalities following application of current shunts. The most common abnormality is tail defects. Tail structures, including neural tube, notochord and somites, are frequently absent or aberrantly formed [12]. Disruption of endogenous electric currents may cause disruption of normal cell migration and organization, as well as growth and differentiation processes, which results in abnormal embryonic development.

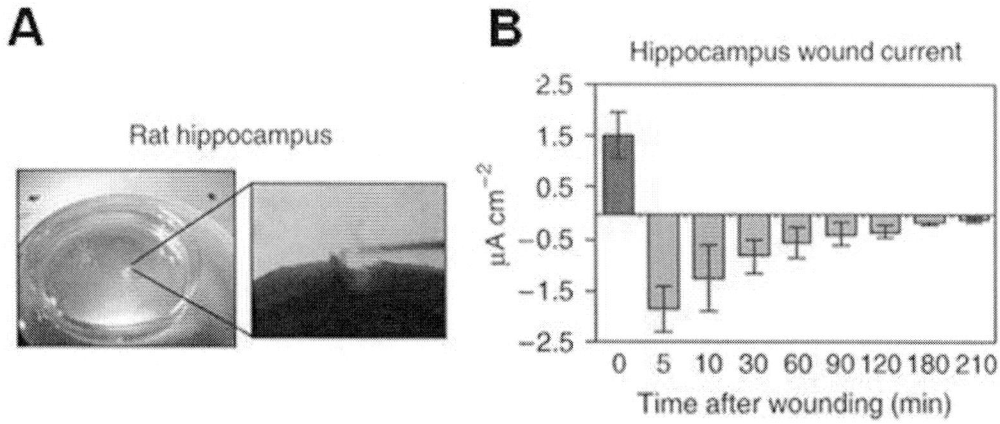

Figure 1. Endogenous current measured in rat hippocampus. A: Rat hippocampus was wounded with fine forceps. Vibrating probe in measuring position near wound. B: Unwounded hippocampal tissue displayed an outward current (red column) and wounding induced a significant inward current (green column). [From Reid et al. [19]]

Damaged to the CNS may induce endogenous EFs [14]. Electrocorticography recording found abnormal cortical electric activity in patients following traumatic brain injury. Cortical spreading depolarizations were recorded in a bimodal pattern with peak incidence on day 1 and 7 [15]. Transection of lamprey spinal cord can generate currents of about 0.5 mA/cm^2 entering the cut surface of the spinal cord. These currents persist for days after injury [16, 17]. In mammalian spinal cord, an enormous endogenous bioelectric current (0.2-0.3 mA/cm^2) transverses the site of injury [18]. We have shown that if the hippocampus is damaged, there are inward electric currents at the site of damage (Figure 1.) [19]. The brain tissue has relatively small electric resistance/impedance, ~tens of ohms [20]. The endogenous EFs thus generated in the brain tissues are likely to be very small, perhaps within a few millivolts. It should therefore be noted, that although applied EFs are able to guide migration of neurons or neural progenitor cells, it is unknown whether the endogenous EFs play a role in neuronal migration.

For cells at skin and corneal wounds, the endogenous EFs are much larger and more likely to play a role. Applied EFs have been shown to guide migration of many types of cells. Because cells at these wounds, especially the epithelial cells, keratinocytes, and corneal epithelial cells, respond to applied EFs of physiological strengths like those measured at wounds, the endogenous wound EFs are likely to play a role in guiding cell migration in epithelial wound healing [21-23]. Scientists are beginning to contemplate guiding neuronal cell migration using EFs following neuronal trauma and neuronal degeneration to aid in treatment, as well as directing neurite growth.

2. Applied EFs Guide Neuronal Migration

Applied EFs have long been shown to guide neurite growth. However, it is only recently that experimental evidence has demonstrated that applied EFs guide migration of neurons and neuronal progenitor cells.

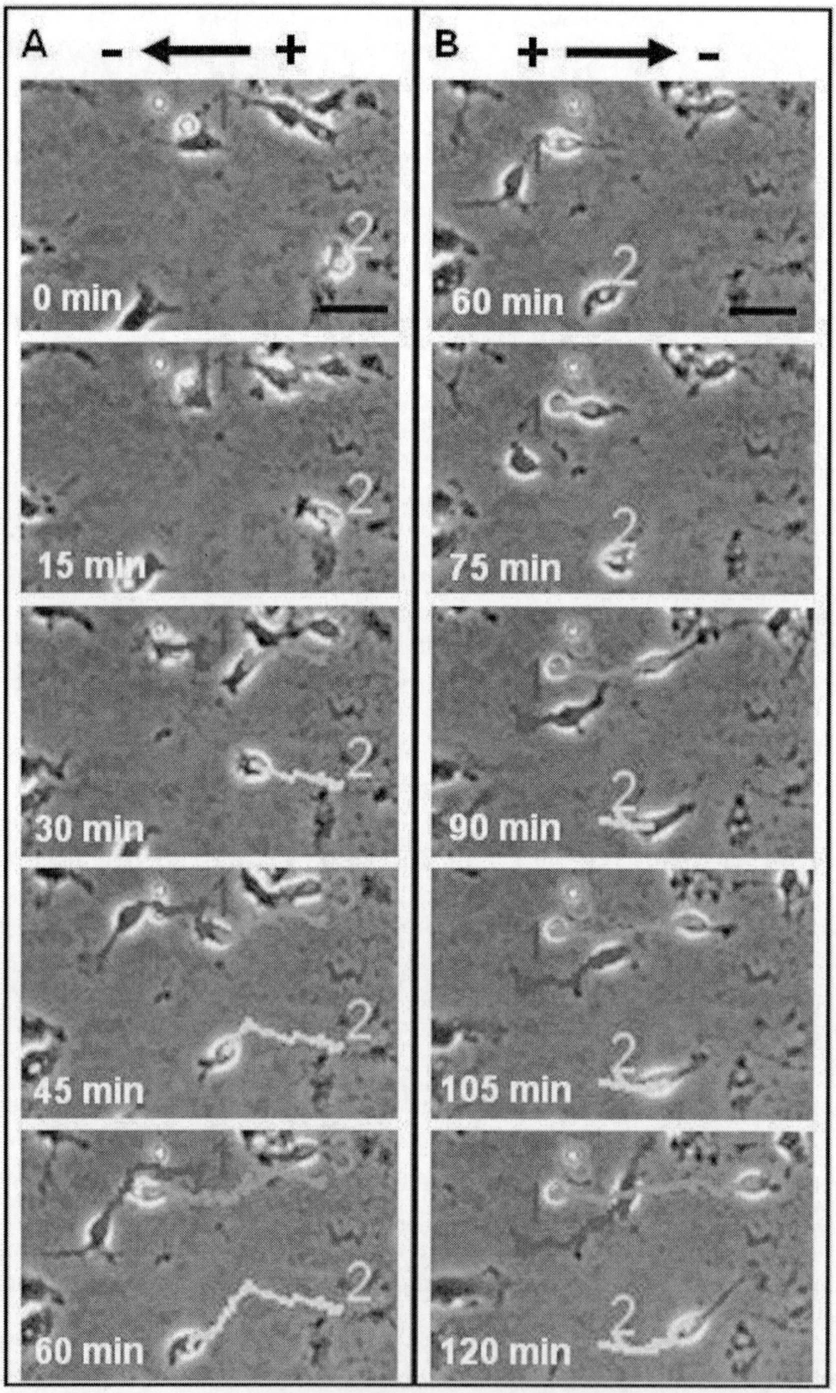

Figure 2. An electric field guides rat hippocampal neurons migration toward the cathode. Time-lapse images of dissociated rat hippocampal neuron migration. A: An electric field with the cathode to the left, guided cells to migrate toward the cathode within 1h. The tracks of neuronal migration are shown by colored trajectories. Cells starting positions are at the numbered ends. B: Reversal of the field polarity reversed the direction of neuronal migration. Scale bar: 50 μm. Field polarity was as shown. EF=150 mV/mm.

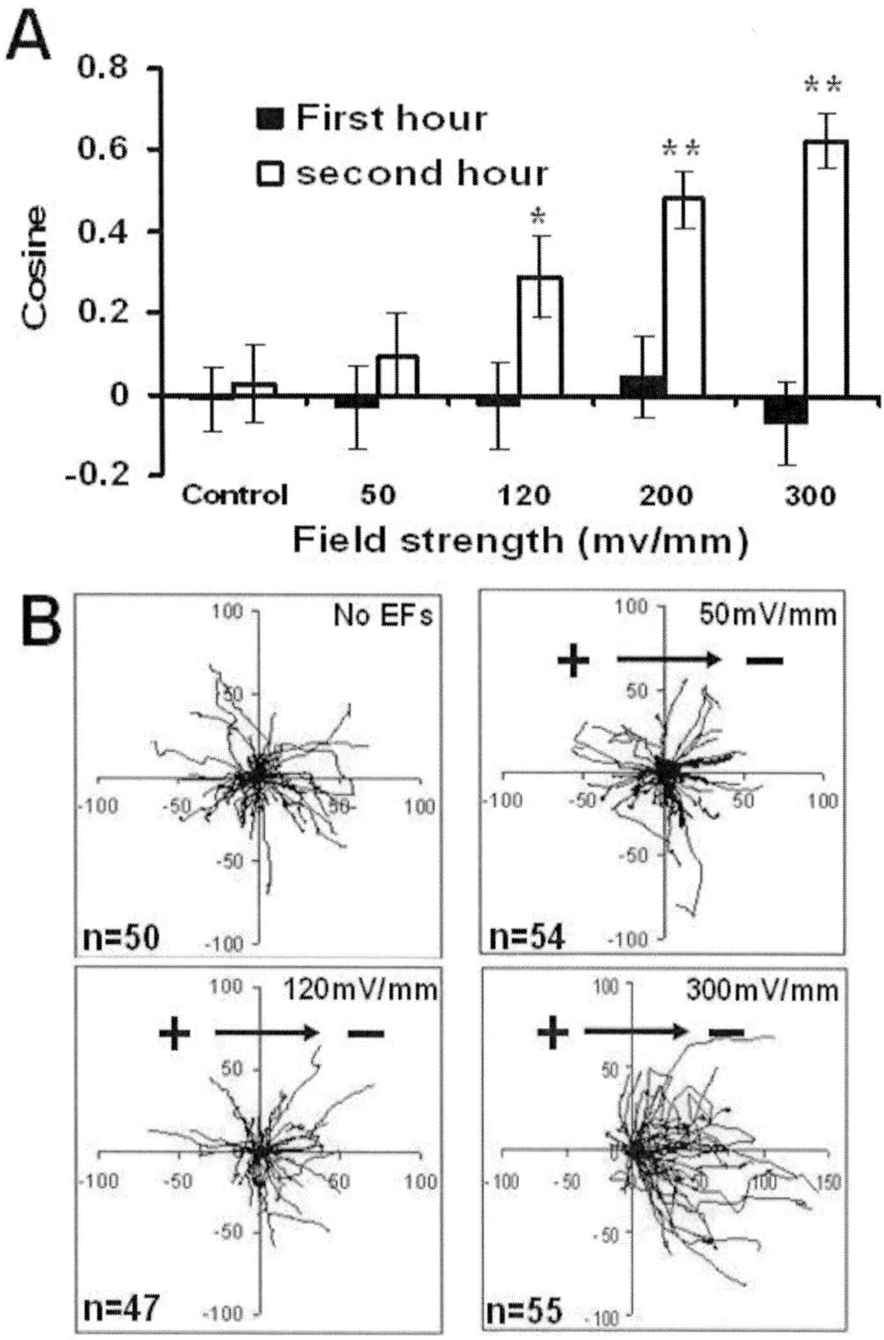

Figure 3. EFs guide rat hippocampal neurons migration toward the cathode in a voltage-dependent manner. A: The directedness (before and after EF stimulation) of rat hippocampal neuron migration in EFs. The average cosine increased with field strength. B: The net translocation of hippocampal neurons during 1h of migration. EFs strengths are indicated in mV/mm. Migration paths were determined by video monitor tracings. The position of all cells at t = 0 min is represented by the origin position (0, 0) with the migratory track of each cell at 60 min plotted as a single line on the graph. The axes are in microns, each arm representing 100 μm of translocation distance. [Modified from Yao et al. [24]]

3.1. Hippocampal Neurons Migrate toward the Cathode

Rat hippocampal neurons respond to direct-current (DC) EFs of 120 mV/mm or above [24]. The cells migrate directionally toward the cathode. When the field polarity is reversed, the cells change their direction of migration and migrate toward the new cathode (Figure 2). This migration was observed both in dissociated cell culture and in explant culture. Cells migrate outward from the explant and in the direction of the cathode. The migration speed of the hippocampal neurons did not change significantly in an electric field.

The directional migration was voltage dependent. In an electric field of 50 mV/mm, the hippocampal neurons did not show directional migration. Applied EFs of 120mV/mm or above were required to induce directional migration (Figure 3). Directedness increased with increasing EFs strength. The directedness of hippocampal neurons reached 0.29 ± 0.10 in an electric field of 120 mV/mm, while it reached 0.62 ± 0.06 in an electric field of 300 mV/mm. These results suggest that the EFs-induced directedness of cathodal migration and displacement, is both time and voltage dependent.

In EFs, hippocampal neurons extended most of their neurites toward the cathode (Figure 4), and the growth cone turned toward the new cathode before neurons reversed their direction of migration toward the new cathode, when the electric field vector was reversed. There are three modes by which neurites reverse their direction of migration; these are to swap, turn, and grow branches of new neurites toward the new cathode. (Figure 4) The most frequently observed pattern is swapping over leading and trailing process [25].

3.2. EFs-guided Migration of Neuronal Progenitor Cells and NSCs

Recently, progress has been made in guiding migration of neural precursors and neural stem cells. This may offer a new technique for stem cell therapies in neuro-trauma and neuronal degenerative disease [26-34]. Applied DC EFs guide the migration of rat embryonic neuronal stem/progenitor cells (NSPCs) toward the cathode. EFs induced NSPCs migrating out from the cathodal side of lateral ganglionic eminence explants to migrate toward the cathode, but prevented the migration of cells on the anodal side of explants toward the anode. The directedness and speed of NSPCs migration toward the cathode was EFs strength-dependent [35]. The EFs increased the migration speed of cells on the cathodal side of the explants migrating toward the cathode. 68% of the cells migrating toward the cathode were nestin- and doublecortin-positive NSPCs. The threshold for directional migration is 50 mV/mm, and the migration speed increased significantly.

Adult rat neural progenitor cells (NPCs) also respond to EFs, with robust directional migration toward the cathode in EFs of 250 mV/mm and 500 mV/mm. EFs guided cell motility, not only *en mass*, but also at the single cell level, toward the cathode, suggesting that EFs regulate cell motility at the single cell level directly, rather than indirectly through modulation of cell-cell interactions. The cell migratory speed increased markedly in EFs, in a voltage-dependent manner [36].

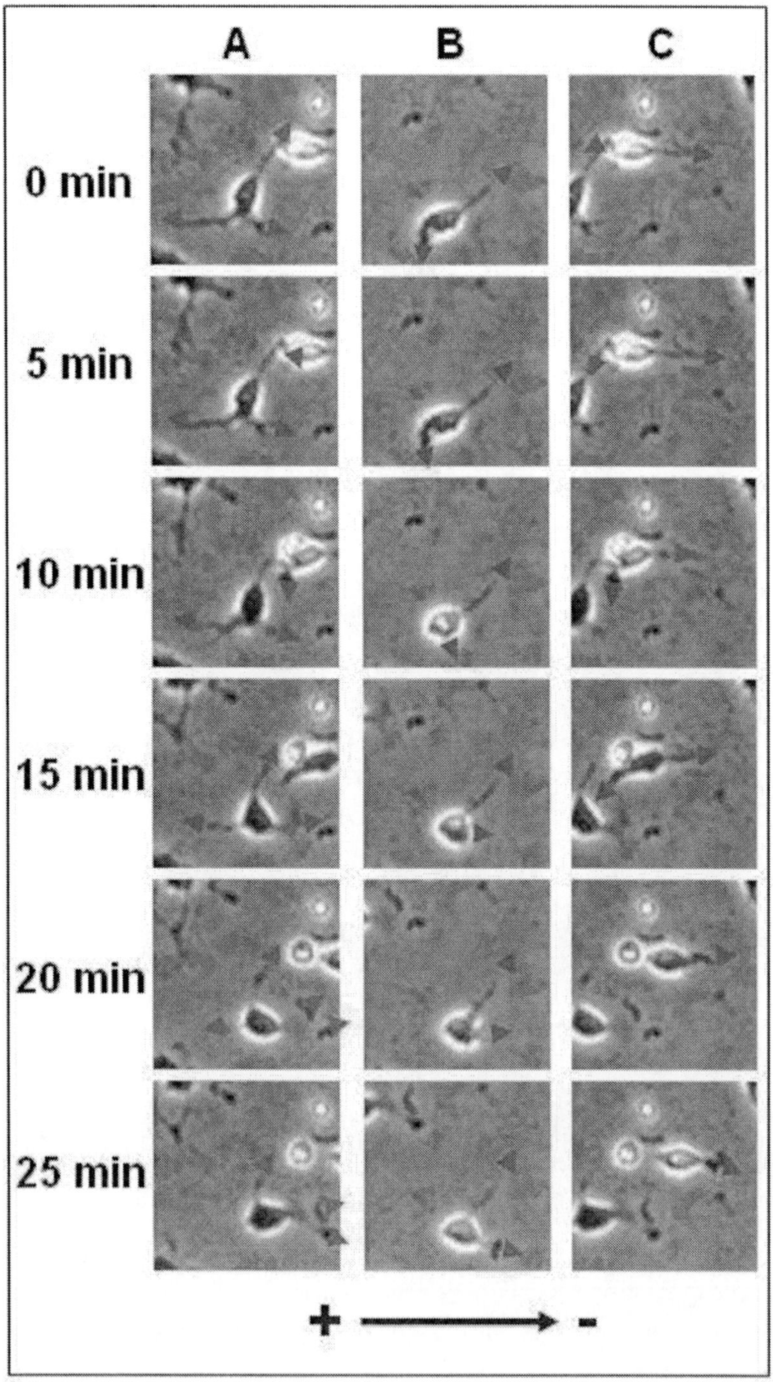

Figure 4. The time-lapse images illustrate leading neurite reorientation in response to EFs steers neuronal migration toward the cathode. There are three ways by which leading processes reorientate in both random migration and EFs-induced migration - trailing processes leading neurite branching (A), neurite turning (B) and swap over of the leading processes (C), red arrows show neurite orientation in EFs. [Modified from Yao et al. [108]]

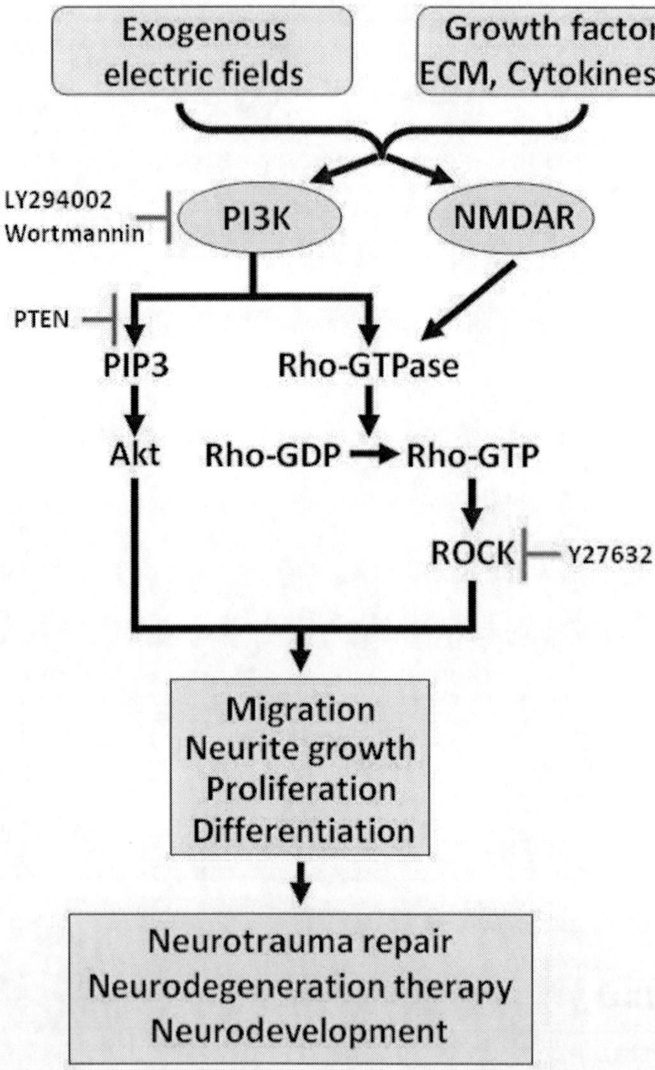

Figure 5. Underlying mechanisms of the effects of EFs on neuronal cell migration. EFs exert multiple effects on neuronal cells. For example, the PI3K/Akt signaling pathway and the NMDAR/Rho GTPase pathway are involved in EFs-induced neuronal cell migration. EFs can cooperate with other guidance cues such as growth factors, cytokines, ECM, etc., to regulate neuronal cell migration, survival, proliferation, and differentiation.

EFs guidance of NSPCs and NPCs migration opens up the possibility to guide transplanted cells to specific sites. Further development of the technique, and optimization of EFs application, may promote great advances in translational medicine.

3.3. Other Types of Nerve Cells

In addition to neurons and neural stem cells, several other types of nerve cells have been found to respond to applied EFs.

Neural crest cells from *Ambystoma mexicanum* and *Xenopus laevis* respond well to EFs. In EFs of more than 5 V/cm, cells migrated toward the cathode [37]. During migration, lateral protrusions were found almost exclusively on the cathode-facing sides of the cell. After 4 days of culture, neural crest cells differentiated into pigment cells. These differentiated pigment cells also responded to applied EFs by becoming perpendicularly aligned. However, these cells oriented slowly, and lateral movement cannot be induced in old cultures. Neural crest cells from 56-h-old quail embryos exhibit directed migration toward the cathode in DC EFs as low as 7 mV/mm. The directedness changed significantly after 7 min of exposure to EFs of more than 10 mV/mm. Moreover, the galvanotactic response was partially reversed by the complete removal of extracellular Ca^{2+}. Both 10 mM Mg^{2+} and 100 μM Gd^{3+} (a Ca^{2+} channel blocker), can inhibit the galvanotactic response. 10 mM Mg^{2+} in Ca^{2+}-free medium totally inhibited the response to EFs, illustrating the fact that the EF guided neural crest cell migration is Ca^{2+}-dependent [38].

Different nerve cell types responded differently to applied EFs. Postganglionic sympathetic neurons (PSNs) retract neurites that are parallel to the EFs vector. Neurites grow out from the soma perpendicular to the lines of force [39]. The orientation effects of EFs on the neurites of rat spiral ganglion neurons varied on different substrates. The growth cones on laminin-coated cultures exposed to steady and pulsed direct current EFs oriented toward the cathode, while those on poly-D-lysine (PDL)-coated cultures oriented toward the anode [40]. Cells from embryonic chick dorsal root ganglia sometimes moved toward the anode [41]. Neuroblasts from the neural tubes of zebrafish embryos were tested in EFs of 100 mV/mm. Neurite growth showed no directional preference in 100 mV/mm EFs. Meanwhile, no growth cone migration was found in EFs of 100 mV/mm [42].

3.4. Neurite Growth in Vitro

Small direct current EFs can promote and orientate neurite growth [43-48]. As early as 1946, Marsh and Beams showed that chick nerve fibers can be guided to grow toward the cathode by electric currents in vitro [49]. Similar results came from embryonic chick dorsal root ganglia. EFs of 70-140 mV/mm induced neurites growing out from explanted embryonic chick ganglia faster toward the cathode than anode. The neurites showed an asymmetrical distribution around micro-explants [41]. In *Xenopus* neurons, EFs seemed to have the effect of inducing neurite outgrowth [50]. When applying EFs of 7-190 mV/mm to single neurons and myoblasts from the neural tube, and somites of *Xenopus laevis* embryos, neurites also grew preferentially toward the cathode [51].

The electric field shaped neurite morphology by increasing the number of filopodia at the growth cone, and by increasing the number of cytoplasmic spines along neurite shafts. Dynamic microfilaments and microtubules play pivotal roles in cathode-directed migration of growth cones in EFs. The microfilament inhibitor latrunculin, the microtubule-stabilizing drug taxol, and the microtubule-destabilizing drugs vinblastine and nocodazole, attenuated the cathodal turning of growth cones in EFs [52]. The EFs induced an asymmetry in the distribution of these cytoplasmic projections; greater numbers were found on the cathodal-facing side than on the anodal-facing side [53]. Concanavalin A (Con A) was found to abolish these effects completely [54].

The effects of EFs on neurite orientation differred on different substrates. Rat spiral ganglion neuron (SGN) neurites turned toward the cathode on laminin-coated slides, whereas they turned toward the anode on PDL-coated slides [40]. Therefore, not all neurites turn toward the cathode in an electric field. Neuron like cells differentiated from PC12 cells orientated their neurites toward the anode [55]. In sharp contrast, isolated embryonic spinal neurons from zebrafish did not show a biased direction of neurite growth at all in an electric field of 100mV/mm, although the level of neurite growth was the same [42, 56].

3.5. Other Cellular Responses of Nerve Cells to EFs

EFs help to weave complicated neuronal network by influencing many different functions of neuronal cell function, in addition to directional migration and neurite growth. Cell orientation, cell growth, cell polarity, cell division, and differentiation all appear to be affected by EFs.

Neuronal cells show altered orientation and morphology in EFs. Neural crest cells from *Ambystoma mexicanum* and *Xenopus laevis* retracted both their anode- and cathode-facing margins and elongated in EFs of 1-5 V/cm [37]. Spherical myoblasts elongated with a bipolar axis of growth perpendicular to the electric field. The threshold of this reaction was 36 mV/mm [51]. An electric field of 437 mV/mm induced perpendicular alignment of adult hippocampal NPCs, their processes to grow perpendicularly to the electric field vector, and inhibited anodally-directed growth of processes [57].

EFs can enhance cell growth and induce differentiation also. Stimulation of *Xenopus* nerve cell growth by applied electric field was observed, and it was found that many more cells sprouted neurites in cultures exposed to EFs, than in control cultures. The numbers of other diffentiating cell types (pigment cells, fibroblast, etc.) also increased. In an electric field of 7 mV/mm, polarity and differentiation of explanted *Xenopus* myoblasts and neuroblasts are influenced. In the absence of EFs, explanted neural tube cells do not differentiate, however, a large fraction do so when exposed to an applied electric field. [51] Adult hippocampal NPCs also show a greater tendency to differentiate into neurons in the presence of an electric field, but not into oligodendrocytes or astrocytes. [57]

EFs appear to be able to orient the direction of cell division of hippocampal cells, both in microtubule-associated protein 2 (MAP-2) positive neuron-like cells, and in glial fibrillary acidic protein (GFAP) positive glial-like cells. The cleavage plane reoriented perpendicularly to the EFs vector, while the mitotic spindle reoriented parallel to the EFs vector. The cell apparatus, such as Golgi, centrosomes, and microtubules, also showed a significant cathodal polarization. [25]

Table 1. Summary of neuronal migration and neurite growth in EFs

Species	Neuronal cell type	Field strength of EF	migration	Neurite growth	Mechanism	References
Xenopus laevis	Neurons and myoblasts	7-190 mV/mm	ND	Grow toward cathode	ND	Hinkle and McCaig. (1981) [51]
Xenopus laevis	Neurons	0.1-10 V/cm	ND	Cathodal facing neurites show accelerated growth	Cathodal accumulation of growth-controlling surface glycoproteins	Patel and Poo. (1982) [54]
Ambystoma mexicanum and Xenopus laevis	Neural crest cells	1-15 V/m	Toward cathode	1-5 V/cm EF retract both anode- and cathode-facing margins 5 V/cm EF induced cathode-facing protrusions formation.	ND	Cooper and Keller. (1984) [37]
Xenopus laevis	Spinal neurites	15-200 mV/mm	ND	Grow or turn toward cathode	ND	McCaig. (1986) [50]
Xenopus laevis	Spinal neurites	30-233 mV/mm	ND	Grow toward cathode	ND	McCaig. (1986) [53]
Xenopus laevis	Earliest developing nerves from dissociated neural tubes	100-150 mV/mm	ND	Promote neurite growth, most field-induced branches arose from cathodal facing side and grew cathodally	Microtubule dynamic	McCaig. (1990) [43]
Xenopus laevis	Embryonic *Xenopus* nerves	120-150 mV/mm	ND	Cathodal galvanotaxis, promote branching and grow faster in EF	ND	McCaig. (1990) [44]

Table 1. Continued

Species	Neuronal cell type	Field strength of EF	migration	Neurite growth	Mechanism	References
Xenopus laevis	Embryonic *Xenopus* nerves	50-200 mV/mm	ND	Induce cathodal turning and cathodal branching of neurites	Modulation of membrane surface charge	Erskine et al. (1995) [46]
Xenopus laevis	Embryonic *Xenopus* nerves	50-200 mV/mm	ND	Turn cathodally	Surface charge with lyotropic anions	Erskine and McCaig. (1995) [47]
Xenopus laevis	Embryonic *Xenopus* nerves	50-133 mV/mm	ND	Cathodal orientation	Activation of growth cone nicotinic AChRs	Erskine and McCaig. (1995) [48]
Xenopus laevis	Neural tube cells	110-120 mV/mm	ND	Neurites growth turn cathode on untreated Falcon tissue culture plastic or on laminin substrata, while turn anode on polylysine	Substrate charge determined	Rajnicek et al. (1998) [109]
Xenopus laevis	Embryonic spinal neurons	150 mV/mm	ND	Growth cone turn toward the cathode	GTPases Cdc42 and Rac signaling	Rajnicek et al. (2006) [52]
Quail	Neural crest cell	0.4-400 mV/mm	Toward cathode	ND	Ca^{2+} signaling	Nuccitelli and Smart. (1989) [38]
Chick	Embryonic nerve fibers	Current density: 0.05-500 μamp./mm^2	ND	Cathodal orientation	ND	Marsh and Beams. (1946) [49]
Chick	Embryonic dorsal root ganglia	70-140 mV/mm	Some times move toward anode	Cathodal facing neurites grow faster than anode.	ND	Jaffe and Poo. (1979) [41]

Chick	Dorsal root ganglia	24 or 44 V/m	ND	Promote neurite growth	ND	Wood and Willits. (2009) [95]
Zebrafish	Embryonic zebrafish neurons	100 mV/mm	No response	No response	ND	Cormie and Robinson. (2007) [42]
Rat	Hippocampal neurons	28, 80, 219 mV/mm	ND	Neurites lay perpendicular to the field, reduce the number of cathode-facing neurites	ND	Rajnicek et al. (1992) [110]
Rat	Hippocampal neurons and explant	50-300 mV/mm	Toward cathode	ND	ROCK and PI3K signaling	Yao et al. (2008) [24]
Rat	Embryonic neuronal stem/progenitor cells (NSPC)	50-250 mV/mm	Toward cathode	ND	NMDAR/Rac1/actin signaling	Li et al. (2008) [35]
Rat	Hippocampal cells (neuron-like cells and glial-like cells)	50-300 mV/mm	Toward cathode	Turning to cathode	ROCK and PI3K signaling	Yao et al. (2009) [25]
Rat	Adult hippocampal neural stem/progenitor cells	DC EF: 437 mV/mm	Toward cathode	Inhibited anodally and promoted perpendicularly	ND	Ariza et al. (2010) [57]
		AC EF: 46 mV/mm	No response	No response	ND	
Rat	Adult neural stem cells	250 mV/mm	Toward cathode	Turn to cathode	PI3K signaling	Arocena et al. (2010) [60]

Table 1. Continued

Species	Neuronal cell type	Field strength of EF	migration	Neurite growth	Mechanism	References
Rat	Spiral ganglion neurons (SGNs)	Steady DC-EF 2.2 V Pulsed DC-EF: 2.2 V, 50 Hz, occupation rate 20% Charge-balanced biphasic puled EF: 250 Hz, pulse width 100 μs	ND	oriented toward the cathode on laminin-coated slides, while those on PDL-coated oriented toward the anode. The neurites in charge-balanced biphasic pulsed EFs had an obvious orientation of turning away from the electrode rings.	ND	Li et al. (2010) [40]
Rat	Embryonic and adult neural progenitor cells (NPCs)	500 mV/mm	Toward cathode	ND	PI3K/Akt signaling	Meng et al. (2011) [36]
Cell line	PC 12 cells	5-100 mV/mm	ND	Neurites grow toward anode	ND	Cork et al. (1994) [55]

3.6. Differences in Electric Field-guided Responses of Neurons from Different Species and Developmental Stages

There are contrasting discrepancies on how different neurons and nerve cells respond to EFs. Both embryonic neural progenitor cells (eNPCs) and adult neural progenitor cells (aNPCs) showed cathodal electrotaxis in EFs of physiological magnitude in vitro. [36] Transplanted NPCs also showed robust directional migration toward the cathode in spinal cord slices ex vivo, when exposed to applied EFs. We reported that hippocampal cells react to EFs of 120 mV/mm and above, while Li et al. reported that NSPCs displayed directional migration in EFs as low as 50 mV/mm. Their results differed from ours in two ways. We used postnatal hippocampal neurons. Their progenitor cells were obtained from the lateral ganglionic eminence of embryonic E17-18 rats. While postnatal development and neuronal migration from the hippocampus remain active, the developmental stage and the differentiation stage may significantly affect the responses to EFs. This may also explain why *Xenopus* spinal neurons respond remarkably well to applied EFs with a very small field strength ~ 7 mV/mm [51], to grow neurites toward the cathode [58], while experiments using neuroblasts from neural tubes of zebrafish embryos however, did not respond to applied EFs of more than 10 times the field strength (100 mV/mm) [42, 59].

Table 1 shows experimental results on neural cell migration and neurite growth in applied EFs.

4. Mechanisms of EFs-directed Neuronal Migration

Why do neuronal cells show directional migration and nerve growth in EFs? How can we evaluate the migration and neurite growth? Researchers have been trying to uncover the mechanisms of EFs-directed neuronal migration for many years. A quantitative model was developed to analyze neural stem cell motion in EFs based on their tracks of movement in EFs. In the absence of EFs, NSCs displayed an amoeboid motion, characterized by small lamellipodial-like protrusions with random and changing orientations, leading to highly tortuous migration. Control of protrusion orientation by EFs is an important component of the electrotactic response. In EFs, tortuosity diminished as electrotaxis toward the cathode occurred. Most protrusions of NSCs oriented toward the cathode. When the phosphoinositide 3-kinase (PI3K) inhibitor LY 294002 was applied, the cells showed an intermediate morphology. Based on these observations, a model of NSC migration was generated with only two key parameters, the probability that a cell will change direction, and the probability that it will choose a particular angle of displacement. The model might be used to predict how variations in its parameters will affect trajectories, and contribute to the understanding of the effect of different treatments on cell migration [60].

Rho GTPase and PI3K signaling are well-studied signaling pathways in cell migration. They are also studied in EFs-guided neuronal cell migration. Suppressing PI3K activation decreases electric field-induced signaling and abolishes directed movements of many cell types that migrate in response to electric signals [61]. Pharmacological inhibition and genetic disruption of PI3K significantly reduced the directional migration of hippocampal neurons and NPCs in EFs, while EFs increased the activity of PIP3 and Akt, which are the effectors of PI3K following stimulation with epidermal growth factor (EGF) and fibroblast growth factor-2 (FGF-2). At the same time, EFs guiding neuronal migration was paralleled with phosphorylated Akt (P-Akt) polarization. P-Akt was observed to concentrate on the cathode-facing side of the cell body when exposed to an electric field. The direction of neurite process extension leads the migratory direction of neurons. Inhibition of the Rho-A effector, Rho-associted coiled-coil containing protein kinase1/2 (ROCK1/2), also completely abolished EFs-guided directional migration. Both PI3K and ROCK1/2 inhibitors lowered the track velocity of neurons, both in the presence and absence of EFs stimulation. Intracellular apparatus polarization was also abolished by PI3K and ROCK1/2 inhibitors. These results suggest that PI3K and ROCK1/2 are involved in EFs-induced neuronal migration and polarization of intracellular apparatus. [24, 25, 62]

EFs also trigger the redistribution of epidermal growth factor receptor (EGFR), phosphatidylinositol (3,4,5)-trisphosphate (PIP3), and actin in NPCs, which showed a cathode-facing asymmetry in EFs. For fibroblast growth factor receptors (FGFR), EFs enhanced the activity and nuclear translocation of FGFR, as visualized by the increased fluorescent intensity in the NPC nucleus following EFs exposure. Withdrawal of EGF and FGF markedly reduced the EFs-induced directional migration of NPCs, demonstrating that the electric field-induced migration of aNPCs and eNPCs, requires the participation of EGF and FGF.

The N-Methyl-D-aspartate receptor (NMDAR)/Rac1/actin signal transduction pathway has been identified to play an important role in EFs guidance of embryonic NSPCs cathodal migration. DAPV, a selective NMDAR antagonist, inhibits EFs-induced NSPCs migration toward the cathode, indicating that activation of NMDAR is required in EFs-induced NSPCs migration. Meanwhile, EFs can enhance the association of NMDARs with Tiam1, Pak1, actin, and the Pak1 phosphorylation, all of which can be inhibited by DAPV inhibition. These results suggest that EFs induce NSPCs cathodal migration by activating NMDARs, and the related Tiam1/Rac1/Pak1/actin pathway [35].

5. Nerve Growth in Vivo

Electric stimulation may also play a critical role in axonal regeneration in vivo as EFs show promising guidance effects in neurite growth and orientation in vitro. Both peripheral nervous system [63] and CNS regeneration have been shown to be enhanced by electric stimulation. The electric field has tropic effects on injured spinal cord axons in in vitro and in vivo models of sea lamprey, rodent, and canine spinal cord injury (SCI), and improves functional outcome.

Electric currents of about 0.5 mA/cm^2 were detected at the surface of the proximal portion of an isolated and transected lamprey spinal cord [16]. Large bioelectric currents were found to enter damaged cells and tissues of guinea pig spinal cords, immediately following damage. This endogenous current declines with time and distance from the local site of injury, but eventually maintains a much lower but stable value ($< 50\ \mu A/cm^2$) [18].

Borgens et al. first applied a weak, steady electric current of approximately 10 microamperes across the completely severed spinal cord of the larval lamprey *Petromyzon marinus* and this was shown to enhance regeneration in the severed giant reticulospinal neurons. In most of the electrically treated animals, process growth from giant axons was observed in, or across, the lesion [64, 65]. The same treatment was performed on severed guinea pig spinal cords. In this case, axons grew into the glial scar, as far as the plane of transection in most experimental animals, following EFs application. Moreover, these fibers returned to their approximate positions within the rostral spinal cord before turning toward the brain; this response was not detected in the control group [66]. An electric field of 200 mV/mm (cathode rostral) applied across a hemisection of guinea pig spinal cord, led to functional recovery of the cutaneous trunci muscle reflex [67, 68]. Further studies showed that this recovery was due to the regeneration of spinal axons around and through the lesion. Long-term (1 month) stimulation across a transection of ascending white matter tracts by a steady, rostrally-negative, weak (0.4 mV/mm) voltage gradient, can guide axons penetrating the lesion, and in some cases branches projected across the plane of transaction [69]. Applying an electric field of 100 mV/mm into a hollow silicone rubber tube within the damaged spinal cord, could guide a robust regeneration of axons into the tube [70]. An electric field of 320 mV/mm across puncture wounds to the adult rat spinal cord induced a slight shift in the orientation of astrocytes, orienting them parallel to the long axis of the spinal cord, and also markedly reduced the relative number of astrocytes, particularly in the lesion [71]. Using an in vivo rat corneal wound model, a steady DC electric field was detected instantaneously at rat corneal wounds. This wound-induced EFs can induce nerves to turn in order to grow perpendicularly toward the wound edge and can stimulate directed nerve sprouting [72].

These findings suggest that an imposed electric field promotes growth of axons following neural trauma, and the possibility that a steady voltage gradient may be an environmental component necessary for axonal development and regeneration.

Electric stimulation (3 V for 0.1 ms at 20 Hz) can increase reinnervation and the number of regenerated myelinated fibers in the distal tibial nerve, suggesting that application of electric stimulation following sciatic nerve injury, promotes axonal regeneration over a long distance, and reduces facilitation of spinal motor responses [73]. A low-frequency electric stimulation method was developed that greatly accelerates axon outgrowth on the proximal nerve stump shortly following surgical repair of a transected peripheral nerve [74]. 20 Hz bipolar electric stimulation was performed for 1 hour to 2 weeks in rats, and in carpal tunnel syndrome (CTS) patients. 1 hour-long electric stimulation accelerated axon outgrowth across the suture site in association with elevated expression of neuronal neurotrophic factor and receptors, and in patients, promoted the full reinnervation of thenar muscles, in contrast to a non-significant increase in motor unit numbers in the control group [75]. Rapid expression of neurotrophic factors and their receptors and cyclic adenosine monophosphate (cAMP)-related growth-associated proteins, is the likely mechanism by which electric stimulation accelerates axon outgrowth from the site of injury in both peripheral and central nervous systems [76]. Recent studies uncovered the role of voltage-gated sodium channels in regeneration. After *Xenopus laevis* tail amputation, increased intracellular Na^+ is required for initiating regeneration. Molecular and pharmacological inhibition of sodium transport causes regenerative failure, while induction of a transient sodium current can restore regeneration [77].

Electric stimulation has been used in clinical trials. A human oscillating field stimulator (OFS) was implanted in 10 patients with complete motor and sensory SCI. The electric stimulation improved light touch, pinprick sensation in most of the patients. In cases of cervical cord injury, bilateral upper-extremity somatosensory evoked potentials (SSEPs) became normal in 1, unilateral upper-extremity SSEPs were recovered in 4, bilateral upper-extremity SSEPs were recovered in 1, and abnormal lower-extremity SSEPs resolved in 1 case following electric stimulation [78].

6. Synergy of EFs and other Guidance Cues for Neuronal Migration

Neuronal migration includes several steps. When migration is initiated, cells transform from a stationary state to a motile form. Migrating neurons extend a short leading process ending in a growth cone similar to those of growing axons. During migration, the cell body follows the growing process [79]. Leading processes of neuronal cells which are similar to the growth cones of migrating axons, are thought to play an important role in sensing the surrounding microenvironment and thereby contributing to the guidance of neurons [80]. Migrating neurons are highly polarized in the direction of their movement. This is achieved through the generation, maintenance, and remodeling of a leading process that marks the direction followed by the cell.

Neuronal cell migration and direction of nerve growth can be regulated by several extracellular cues, including growth factors, extracellular matrix and other soluble and substrate bound molecules [79, 81-87]. The interaction of EFs and other guidance cues orchestrate the formation and regulation of the nervous system.

Chemotaxis is the major mechanism thought to direct neuronal migration [88]. The gradients of chemicals may attract nerve growth, orchestrating cell migration. Interneurons respond to chemoattractant signals by generating new leading process branches that are better aligned with the source of the gradient [89]. Concentration gradient of molecular cues, such as the secreted protein Slit, guides the direction of migrating neurons and axon projection by interacting with the Rho-GTPase family of proteins in vivo [90-92]. Focal adhesion kinase (FAK) and Src family kinases were found to be necessary in axon outgrowth and attraction [93]. Chemokine stromal-derived factor 1 (SDF-1) functions as an attractive guidance cue for neuronal migration, which is important for controlling the migration of embryonic external germinal layer (EGL) cells [94]. EFs stimulation with N2 supplement and nerve growth factor (NGF) resulted in 47% and 44% longer neurite growth respectively, compared to controls [95]. Other growth factors, like EGF and FGF also promote neuronal migration in an electric field. Withdrawal of EGF and FGF markedly reduced the EFs-induced directional migration of NPCs illustrating the fact that electric field-induced aNPC and eNPCs migration requires the participation of EGF and FGF. [36]

Extracellular matrix (ECM) molecules shown to be involved in neuronal migration include reelin [96], integrins [97], heparin sulfate proteoglycans (HSPGs) [98], laminin [99, 100], and anosmin-1 [101]. The physical nature of the substrate can also determine the direction of nerve growth. Studies have shown that ECM may influence neuronal migration by contributing to the formation of gradients of attractants or repellents, by allowing for the combined presentation of some regulatory signals, and by modulating or potentiating the signaling of receptors for guidance or migration modulating cues. The effects of EFs on neurite orientation differ on different substrates. Rat SGN neurites turned toward the cathode on laminin-coated slides while they turned toward the anode on PDL-coated slides [40]. Laminin, instead of collagen, can enhance dorsal root ganglion neurite growth length when combined with EFs of 24 V/m or 44 V/m [95].

Cell adhesion molecules (CAMs) and related cell surface proteins also play a pivotal role in axon pathfinding and neuronal migration. Poly-sialated neural cell adhesion molecule (PSA-NCAM) is a critical regulator of neuronal precursor migration through the rostral migratory stream, in which neuroblasts migrate from the subventricular zone to the olfactory bulb [102-104].

Topography is another factor which regulates directional migration of neuronal cells. Neuronal PC12 cells migrate directionally on an anisotropic and periodic collagen-coated area between the ratchet wheel-shaped outer frame and round inner frame of polydimethylsiloxane [105]. Topographical information can also be provided by generating gradients in expression levels of membrane bound factors and/or receptors. Glial cells may also contribute to the initiation and modulation of movement of neuronal precursors through the release of a diffusible protein [106]. Combination of EFs and parallel substratum grooves guide neurite growth synergistically. The *Xenopus* neurons cathodal migration was enhanced when the EFs vector and the groove axis were presented simultaneously and parallel to each other. If the EFs vector was presented at a right angle to the groove axis, the response of neuronal growth cones depended on groove width. Growth cones migrated along grooves of 1 or 2 μm width and migrated toward the cathode on 4 μm grooves. [107]

Conclusion and Perspectives

Endogenous EFs are present in the nervous system during development. Damage or injury to the adult nervous system may produce very small endogenous EFs. Very limited information is available on the magnitude and distribution of these endogenous EFs however. Whether these endogenous EFs are merely due to a passive leakage of ions and other charged molecules from the damaged cells, or cells and tissues actively participate in the generation and regulation of the fields is still debated. Development of new techniques may help answer this question. Further research is required to explore the map of electric potentials in the mammalian brain, which may be an important factor in the development and repair of the nervous system.

Externally applied EFs are able to guide migration of neurons and NSCs, growth cones, and growth of nerve fibers. The effects are evident and very strong in the majority of cell types studied, including rat and mouse neurons, and NSCs. However, striking differences between species and nerve cell types are an important consideration. Further understanding of the mechanisms underlying EFs-induced cellular responses may shed light on these differences.

The strong effects of applied EFs on the guidance of neuronal migration and nerve growth suggest a promising novel modality in nerve engineering and nerve repair. Figure 5 illustrates a simplified description of the pathways that transduce the electric signal to nerve repair through guiding neuronal migration and neurite growth, and which may offer promising novel modalities to the treatment of neurotrauma and degeneration. One of the most challenging issues in using EFs in the human body/nervous system as a therapy, is that live tissues are volume conductors and are electrically heterogeneous in resistance and therefore in field distribution. Maintaining steady and effective fields could be challenging. Studies in animal models therefore, are critical to the development of plausible techniques for applying EFs, and testing effects in vivo will be the next challenge in the field. Only with this as a basis, can we explore further therapies in neurotrauma and neurodegenerative disease through combination of electric stimulation, cell transplantation, pharmacological treatment, growth factors, cytokines, and extracellular modulation.

ACKNOWLEDGMENTS

This work is supported by grants from the California Institute of Regenerative Medicine RB1-01417 (to MZ). MZ is also supported by NIH 1R01EY019101, NSF MCB-0951199, and UC Davis Dermatology Developmental Fund, and in part by the Research to Prevent Blindness, Inc. LZ is supported partially by National Natural Science Foundation of China 30900274 (to LZ) and 30671738. LZ and ZY are supported by National 973 Project Grant of China 2011CB503700 (to ZY). We thank Zhao lab members for helps, and Dr. Laura Chalmers for editorial help.

REFERENCES

[1] Magavi SS, Leavitt BR, Macklis JD. Induction of neurogenesis in the neocortex of adult mice. *Nature.* 2000;405:951-955.

[2] Nakatomi H, Kuriu T, Okabe S, et al. Regeneration of hippocampal pyramidal neurons after ischemic brain injury by recruitment of endogenous neural progenitors. *Cell.* 2002;110:429-441.

[3] O'Donnell M, Chance RK, Bashaw GJ. Axon growth and guidance: receptor regulation and signal transduction. *Annu. Rev. Neurosci.* 2009;32:383-412.

[4] Sawamoto K, Wichterle H, Gonzalez-Perez O, et al. New neurons follow the flow of cerebrospinal fluid in the adult brain. *Science.* 2006;311:629-632.

[5] Hotary KB, Robinson KR. The neural tube of the Xenopus embryo maintains a potential difference across itself. *Brain Res Dev Brain Res.* 1991;59:65-73.

[6] Hotary KB, Robinson KR. Endogenous electric currents and voltage gradients in Xenopus embryos and the consequences of their disruption. *Dev Biol.* 1994;166:789-800.

[7] Robinson KR, Stump RF. Self-generated electric currents through Xenopus neurulae. *J Physiol.* 1984;352:339-352.

[8] Shi R, Borgens RB. Embryonic neuroepithelial sodium transport, the resulting physiological potential, and cranial development. *Dev Biol.* 1994;165:105-116.

[9] Shi R, Borgens RB. Three-dimensional gradients of voltage during development of the nervous system as invisible coordinates for the establishment of embryonic pattern. *Dev Dyn.* 1995;202:101-114.

[10] Jaffe LF, Stern CD. Strong electric currents leave the primitive streak of chick embryos. *Science.* 1979;206:569-571.

[11] Hotary KB, Robinson KR. Endogenous electric currents and the resultant voltage gradients in the chick embryo. *Dev Biol.* 1990;140:149-160.

[12] Hotary KB, Robinson KR. Evidence of a role for endogenous electric fields in chick embryo development. *Development.* 1992;114:985-996.

[13] Erskine L, Stewart R, McCaig CD. Electric field-directed growth and branching of cultured frog nerves: effects of aminoglycosides and polycations. *J Neurobiol.* 1995;26:523-536.

[14] Borgens RB. Stimulation of neuronal regeneration and development by steady electric fields. *Adv. Neurol.* 1988;47:547-564.

[15] Hartings JA, Strong AJ, Fabricius M, et al. Spreading depolarizations and late secondary insults after traumatic brain injury. *J. Neurotrauma.* 2009;26:1857-1866.

[16] Borgens RB, Jaffe LF, Cohen MJ. Large and persistent electric currents enter the transected lamprey spinal cord. *Proc. Natl. Acad. Sci. U. S. A.* 1980;77:1209-1213.

[17] Borgens RB. Voltage gradients and ionic currents in injured and regenerating axons. *Adv. Neurol.* 1988;47:51-66.

[18] Zuberi M, Liu-Snyder P, Ul Haque A, et al. Large naturally-produced electric currents and voltage traverse damaged mammalian spinal cord. *J Biol Eng.* 2008;2:17.

[19] Reid B, Nuccitelli R, Zhao M. Non-invasive measurement of bioelectric currents with a vibrating probe. *Nat Protoc.* 2007;2:661-669.

[20] Harting MT, Smith CT, Radhakrishnan RS, et al. Regional differences in cerebral edema after traumatic brain injury identified by impedance analysis. *J Surg Res.* 2010;159:557-564.

[21] Zhao M. Electric fields in wound healing-An overriding signal that directs cell migration. *Semin. Cell Dev. Biol.* 2009;20:674-682.

[22] McCaig CD, Rajnicek AM, Song B, et al. Controlling cell behavior electrically: current views and future potential. *Physiol. Rev.* 2005;85:943-978.

[23] McCaig CD, Song B, Rajnicek AM. Electric dimensions in cell science. *J Cell Sci.* 2009;122:4267-4276.

[24] Yao L, Shanley L, Mccaig C, et al. Small applied EFs guide migration of hippocampal neurons. *J. Cell. Physiol.* 2008;216:527-535.

[25] Yao L, McCaig CD, Zhao M. Electric signals polarize neuronal organelles, direct neuron migration, and orient cell division. *Hippocampus.* 2009;19:855-868.

[26] Svendsen CN, Caldwell MA. Neural stem cells in the developing central nervous system: implications for cell therapy through transplantation. *Prog Brain Res.* 2000;127:13-34.

[27] Brock SC, Bonsall J, Luskin MB. The neuronal progenitor cells of the forebrain subventricular zone: intrinsic properties in vitro and following transplantation. *Methods.* 1998;16:268-281.

[28] Chen K, Hughes SM, Connor B. Neural progenitor cells derived from the adult rat subventricular zone: characterization and transplantation. *Cell Transplant.* 2007;16:799-810.

[29] Hicks AU, MacLellan CL, Chernenko GA, et al. Long-term assessment of enriched housing and subventricular zone derived cell transplantation after focal ischemia in rats. *Brain Res.* 2008;1231:103-112.

[30] Madhavan L, Daley BF, Paumier KL, et al. Transplantation of subventricular zone neural precursors induces an endogenous precursor cell response in a rat model of Parkinson's disease. *J Comp Neurol.* 2009;515:102-115.

[31] Mignon L, Vourc'h P, Romero-Ramos M, et al. Transplantation of multipotent cells extracted from adult skeletal muscles into the subventricular zone of adult rats. *J Comp Neurol.* 2005;491:96-108.

[32] Mligiliche NL, Xu Y, Matsumoto N, et al. Survival of neural progenitor cells from the subventricular zone of the adult rat after transplantation into the host spinal cord of the same strain of adult rat. *Anat Sci Int.* 2005;80:229-234.

[33] Zhang P, Li J, Liu Y, et al. Human embryonic neural stem cell transplantation increases subventricular zone cell proliferation and promotes peri-infarct angiogenesis after focal cerebral ischemia. *Neuropathology.* 2010.

[34] Zigova T, Pencea V, Betarbet R, et al. Neuronal progenitor cells of the neonatal subventricular zone differentiate and disperse following transplantation into the adult rat striatum. *Cell Transplant.* 1998;7:137-156.

[35] Li L, El-Hayek YH, Liu B, et al. Direct-current electric field guides neuronal stem/progenitor cell migration. *Stem Cells.* 2008;26:2193-2200.

[36] Meng X, Arocena M, Penninger J, et al. PI3K mediated electrotaxis of embryonic and adult neural progenitor cells in the presence of growth factors. *Exp Neurol.* 2011;227:210-217.

[37] Cooper MS, Keller RE. Perpendicular orientation and directional migration of amphibian neural crest cells in dc electric fields. *Proc. Natl. Acad. Sci. U. S. A.* 1984;81:160-164.

[38] Nuccitelli R, Smart T. Extracellular Calcium Levels Strongly Influence Neural Crest Cell Galvanotaxis. *Biol. Bull.* 1989;176:130-135.

[39] Pan L, Borgens RB. Perpendicular organization of sympathetic neurons within a required physiological voltage. *Exp Neurol.* 2010;222:161-164.

[40] Li S, Li H, Wang Z. Orientation of spiral ganglion neurite extension in electric fields of charge-balanced biphasic pulses and direct current in vitro. *Hear Res.* 2010; 267: 111-118.

[41] Jaffe LF, Poo MM. Neurites grow faster toward the cathode than the anode in a steady field. *J Exp Zool.* 1979;209:115-128.

[42] Cormie P, Robinson KR. Embryonic zebrafish neuronal growth is not affected by an applied electric field in vitro. *Neurosci Lett.* 2007;411:128-132.

[43] McCaig CD. Nerve branching is induced and oriented by a small applied electric field. *J. Cell Sci.* 1990;95 (Pt 4):605-615.

[44] McCaig CD. Nerve growth in a small applied electric field and the effects of pharmacological agents on rate and orientation. *J. Cell Sci.* 1990;95 (Pt 4):617-622.

[45] McCaig CD. Nerve guidance: a role for bio-EFs? Prog Neurobiol. 1988;30:449-468.

[46] Erskine L, Stewart R, McCaig CD. Electric field-directed growth and branching of cultured frog nerves: effects of aminoglycosides and polycations. *J. Neurobiol.* 1995;26:523-536.

[47] Erskine L, McCaig CD. The effects of lyotropic anions on electric field-induced guidance of cultured frog nerves. *The Journal of physiology.* 1995;486 (Pt 1):229-236.

[48] Erskine L, McCaig CD. Growth cone neurotransmitter receptor activation modulates electric field-guided nerve growth. *Dev Biol.* 1995;171:330-339.

[49] Marsh G, Beams HW. In vitro control of growing chick nerve fibers by applied electric currents. *J. Cell. Physiol.* 1946;27:139-157.
[50] McCaig CD. EFs, contact guidance and the direction of nerve growth. *J. Embryol. Exp. Morphol.* 1986;94:245-255.
[51] Hinkle L, McCaig CD, Robinson KR. The direction of growth of differentiating neurones and myoblasts from frog embryos in an applied electric field. *J Physiol.* 1981;314:121-135.
[52] Rajnicek AM, Foubister LE, McCaig CD. Growth cone steering by a physiological electric field requires dynamic microtubules, microfilaments and Rac-mediated filopodial asymmetry. *J Cell Sci.* 2006;119:1736-1745.
[53] McCaig CD. Dynamic aspects of amphibian neurite growth and the effects of an applied electric field. *The Journal of physiology.* 1986;375:55-69.
[54] Patel N, Poo MM. Orientation of neurite growth by extracellular EFs. *J Neurosci.* 1982;2:483-496.
[55] Cork RJ, McGinnis ME, Tsai J, et al. The growth of PC12 neurites is biased toward the anode of an applied electric field. *J. Neurobiol.* 1994;25:1509-1516.
[56] Robinson KR, Cormie P. Electric field effects on human spinal injury: Is there a basis in the in vitro studies? *Dev Neurobiol.* 2008;68:274-280.
[57] Ariza CA, Fleury AT, Tormos CJ, et al. The influence of EFs on hippocampal neural progenitor cells. *Stem cell reviews.* 2010;6:585-600.
[58] McCaig CD, Zhao M. Physiological electric fields modify cell behaviour. *Bioessays.* 1997;19:819-826.
[59] Robinson KR, Cormie P. Electric field effects on human spinal injury: Is there a basis in the in vitro studies? *Developmental neurobiology.* 2008;68:274-280.
[60] Arocena M, Zhao M, Collinson JM, et al. A time-lapse and quantitative modelling analysis of neural stem cell motion in the absence of directional cues and in EFs. *J Neurosci Res.* 2010;88:3267-3274.
[61] Zhao M, Song B, Pu J, et al. Electric signals control wound healing through phosphatidylinositol-3-OH kinase-gamma and PTEN. *Nature.* 2006;442:457-460.
[62] Yao L, Pandit A, Yao S, et al. Electric Field-Guided Neuron Migration: A Novel Approach in Neurogenesis. *Tissue Eng Part B Rev.* 2011.
[63] Xu QG, Singh B, Gordon T, et al. Electric Stimulation Augments Peripheral Nerve Regeneration through Sciatic Nerve Transection Injury Gaps. *J. Peripher. Nerv. Syst.* 2009;14:159-159.
[64] Borgens RB, Roederer E, Cohen MJ. Enhanced spinal cord regeneration in lamprey by applied EFs. *Science.* 1981;213:611-617.
[65] Borgens RB, Blight AR, Murphy DJ. Axonal regeneration in spinal cord injury: a perspective and new technique. *J Comp Neurol.* 1986;250:157-167.
[66] Borgens RB, Blight AR, Murphy DJ, et al. Transected dorsal column axons within the guinea pig spinal cord regenerate in the presence of an applied electric field. *J Comp Neurol.* 1986;250:168-180.
[67] Borgens RB, Blight AR, McGinnis ME. Behavioral recovery induced by applied EFs after spinal cord hemisection in guinea pig. *Science.* 1987;238:366-369.
[68] Borgens RB, Blight AR, McGinnis ME. Functional recovery after spinal cord hemisection in guinea pigs: the effects of applied EFs. *J Comp Neurol.* 1990;296:634-653.

[69] Borgens RB, Bohnert DM. The responses of mammalian spinal axons to an applied DC voltage gradient. *Exp Neurol.* 1997;145:376-389.

[70] Borgens RB. Electrically mediated regeneration and guidance of adult mammalian spinal axons into polymeric channels. *Neuroscience.* 1999;91:251-264.

[71] Moriarty LJ, Borgens RB. An oscillating extracellular voltage gradient reduces the density and influences the orientation of astrocytes in injured mammalian spinal cord. *J Neurocytol.* 2001;30:45-57.

[72] Song B, Zhao M, Forrester J, et al. Nerve regeneration and wound healing are stimulated and directed by an endogenous electric field in vivo. *J. Cell Sci.* 2004;117:4681-4690.

[73] Vivo M, Puigdemasa A, Casals L, et al. Immediate electric stimulation enhances regeneration and reinnervation and modulates spinal plastic changes after sciatic nerve injury and repair. *Exp. Neurol.* 2008;211:180-193.

[74] Gordon T, Sulaiman OAR, Ladak A. Electric Stimulation for Improving Nerve Regeneration: Where Do We Stand? *Essays on Peripheral Nerve Repair and Regeneration.* 2009;87:433-444.

[75] Gordon T, Brushart TM, Chan KM. Augmenting nerve regeneration with electric stimulation. *Neurol. Res.* 2008;30:1012-1022.

[76] Gordon T, Udina E, Verge VMK, et al. Brief Electric Stimulation Accelerates Axon Regeneration in the Peripheral Nervous System and Promotes Sensory Axon Regeneration in the Central Nervous System. *Motor Control.* 2009;13:412-441.

[77] Tseng AS, Beane WS, Lemire JM, et al. Induction of vertebrate regeneration by a transient sodium current. *J Neurosci.* 2010;30:13192-13200.

[78] Shapiro S, Borgens R, Pascuzzi R, et al. Oscillating field stimulation for complete spinal cord injury in humans: a phase 1 trial. *J Neurosurg Spine.* 2005;2:3-10.

[79] Song H, Poo M. The cell biology of neuronal navigation. *Nat Cell Biol.* 2001;3:E81-88.

[80] Rakic P. Principles of neural cell migration. *Experientia.* 1990;46:882-891.

[81] Gipson IK, Inatomi T. Extracellular matrix and growth factors in corneal wound healing. *Curr. Opin. Ophthalmol.* 1995;6:3-10.

[82] Sutherland D, Samakovlis C, Krasnow MA. branchless encodes a Drosophila FGF homolog that controls tracheal cell migration and the pattern of branching. *Cell.* 1996;87:1091-1101.

[83] Zheng C, Heintz N, Hatten ME. CNS gene encoding astrotactin, which supports neuronal migration along glial fibers. *Science* (New York, N.Y. 1996;272:417-419.

[84] Mueller BK. Growth cone guidance: first steps toward a deeper understanding. *Annu Rev Neurosci.* 1999;22:351-388.

[85] Song HJ, Poo MM. Signal transduction underlying growth cone guidance by diffusible factors. *Curr Opin Neurobiol.* 1999;9:355-363.

[86] Duchek P, Rorth P. Guidance of cell migration by EGF receptor signaling during Drosophila oogenesis. *Science.* 2001;291:131-133.

[87] Duchek P, Somogyi K, Jekely G, et al. Guidance of cell migration by the Drosophila PDGF/VEGF receptor. *Cell.* 2001;107:17-26.

[88] Ayala R, Shu T, Tsai LH. Trekking across the brain: the journey of neuronal migration. *Cell.* 2007;128:29-43.

[89] Martini FJ, Valiente M, Lopez Bendito G, et al. Biased selection of leading process branches mediates chemotaxis during tangential neuronal migration. *Development.* 2009;136:41-50.

[90] Wu W, Wong K, Chen J, et al. Directional guidance of neuronal migration in the olfactory system by the protein Slit. *Nature.* 1999;400:331-336.

[91] Wong K, Ren XR, Huang YZ, et al. Signal transduction in neuronal migration: roles of GTPase activating proteins and the small GTPase Cdc42 in the Slit-Robo pathway. *Cell.* 2001;107:209-221.

[92] Li HS, Chen JH, Wu W, et al. Vertebrate slit, a secreted ligand for the transmembrane protein roundabout, is a repellent for olfactory bulb axons. *Cell.* 1999;96:807-818.

[93] Liu G, Beggs H, Jurgensen C, et al. Netrin requires focal adhesion kinase and Src family kinases for axon outgrowth and attraction. *Nat Neurosci.* 2004;7:1222-1232.

[94] Zhu Y, Yu T, Zhang XC, et al. Role of the chemokine SDF-1 as the meningeal attractant for embryonic cerebellar neurons. *Nat Neurosci.* 2002;5:719-720.

[95] Wood MD, Willits RK. Applied electric field enhances DRG neurite growth: influence of stimulation media, surface coating and growth supplements. *J Neural Eng.* 2009;6:046003.

[96] Rice DS, Curran T. Role of the reelin signaling pathway in central nervous system development. *Annu Rev Neurosci.* 2001;24:1005-1039.

[97] Jacques TS, Relvas JB, Nishimura S, et al. Neural precursor cell chain migration and division are regulated through different beta1 integrins. *Development.* 1998;125:3167-3177.

[98] Hu H. Cell-surface heparan sulfate is involved in the repulsive guidance activities of Slit2 protein. *Nat Neurosci.* 2001;4:695-701.

[99] Liesi P. Neuronal migration on laminin involves neuronal contact formation followed by nuclear movement inside a preformed process. *Exp. Neurol.* 1992;117:103-113.

[100] Liesi P, Hager G, Dodt HU, et al. Domain-specific antibodies against the B2 chain of laminin inhibit neuronal migration in the neonatal rat cerebellum. *J. Neurosci. Res.* 1995;40:199-206.

[101] Soussi-Yanicostas N, Faivre-Sarrailh C, Hardelin JP, et al. Anosmin-1 underlying the X chromosome-linked Kallmann syndrome is an adhesion molecule that can modulate neurite growth in a cell-type specific manner. *J. Cell Sci.* 1998;111 (Pt 19):2953-2965.

[102] Cremer H, Lange R, Christoph A, et al. Inactivation of the N-CAM gene in mice results in size reduction of the olfactory bulb and deficits in spatial learning. *Nature.* 1994;367:455-459.

[103] Ono K, Tomasiewicz H, Magnuson T, et al. N-CAM mutation inhibits tangential neuronal migration and is phenocopied by enzymatic removal of polysialic acid. *Neuron.* 1994;13:595-609.

[104] Tomasiewicz H, Ono K, Yee D, et al. Genetic deletion of a neural cell adhesion molecule variant (N-CAM-180) produces distinct defects in the central nervous system. *Neuron.* 1993;11:1163-1174.

[105] Ohnuma K, Toyota T, Ariizumi T, et al. Directional migration of neuronal PC12 cells in a ratchet wheel shaped microchamber. *J Biosci Bioeng.* 2009;108:76-83.

[106] Mason HA, Ito S, Corfas G. Extracellular signals that regulate the tangential migration of olfactory bulb neuronal precursors: inducers, inhibitors, and repellents. *The Journal of neuroscience : the official journal of the Society for Neuroscience.* 2001;21:7654-7663.

[107] Rajnicek AM. Directing cell shape and migration by topographic and electric signals. *European Cells and Materials.* 2002;4:2.

[108] Yao L, Pandit A, Yao S, et al. Ele*ctric Field-Guided Neuron Migration: A Novel Approach in Neurogenesis. Tissue engineering.* Part B, Reviews. 2011.

[109] Rajnicek AM, Robinson KR, McCaig CD. The direction of neurite growth in a weak DC electric field depends on the substratum: contributions of adhesivity and net surface charge. *Dev Biol.* 1998;203:412-423.

[110] Rajnicek AM, Gow NA, McCaig CD. Electric field-induced orientation of rat hippocampal neurones in vitro. *Exp. Physiol.* 1992;77:229-232.

In: Neuronal Migration
Editors: Andrew Girard and Luca Moreau
ISBN: 978-1-62100-059-4

Chapter IV

Membrane-Cytoskeletal Interactions in the Regulation of Neuronal Migration in the Adult Brain

Robert D. Beech*
Yale University, Department of Psychiatry,
New Haven, Connecticut, US

Abstract

At least two areas of the adult mammalian brain continue to produce new neurons throughout the adult lifespan: the hippocampus and the olfactory bulb. These new neurons are derived from stem cells located in the subgranular zone of the hippocampus, and the anterior portion subventricular zone surrounding the lateral ventricles, respectively. In the hippocampus, precursor cells divide in the subgranular zone and migrate relatively short distances into the overlying granule cell layer, where they differentiate into new granule cell neurons. In contrast, neuronal precursors destined for the olfactory bulb must migrate long distances first tangentially through the rostral migratory stream, and then radially into the outer layers of the olfactory bulb where they differentiate into two types of inhibitory interneurons: granule cells, and periglomerular cells. This migration requires an ongoing rearrangement of the cytoskeleton as well as integration of chemo-attractive and repulsive cues from the environment to result in directed movement of the cell along the appropriate pathways. Molecules proposed to play a role in the migration of neuronal precursors along the rostral migratory stream include polysialated neural cell adhesion molecule, netrins, semaphorins, slits, ephrins, and neuregulins. In addition, two molecules in the extracellular matrix, reelin and Tenascin-R, as well as the secreted factor prokineticin-2 have been proposed to key roles in the transition from radial to tangential migration. In this review, I discuss what is known about the factors regulating the migration of adult neuronal precursors to the

* Correspondence concerning this article should be addressed to Robert D. Beech, MD, Ph.D. Assistant Professor Yale University, Department of Psychiatry, 300 George St., 8304C, New Haven, CT 06511.

olfactory bulb, as well as recent unpublished data from our laboratory suggesting possible roles of members of the ERM (ezrin-radixin-moesin), cadherin, and catenin families in this process.

Keywords: Cell migration, subventricular zone, rostral migratory stream, polysialated neural cell adhesion molecule, netrin, semaphorin, slits, ephrin, neuregulin, extracellular matrix, reelin, Tenascin-R, prokineticin-2, ERM (ezrin-radixin-moesin), cadherin, and catenin

INTRODUCTION

While many areas of the adult CNS contain stem cells that are capable of giving rise to neurons *in vitro,* only stem cells located in the hippocampal dentate gyrus and the subventricular zone appear to give rise to new adult neurons *in vivo*. This is thought to be the result of bi-directional communication between multipotent stem cells and the three-dimensional local milieu or niche that they inhabit (Cheng et al., 2005b; Spradling et al., 2001; Wurmser et al., 2004). The concept of a unique niche that serves to maintain the "stemness" of progenitor cells was first proposed for hematopoetic progenitors in the late 1960's (Wolf and Trentin, 1968), and more recently has been applied to progenitors of adult generated neurons (Palmer et al., 2000). In vitro experiments have indicated important roles for both glial (Lim and Alvarez-Buylla, 1999; Song et al., 2002) and endothelial (Shen et al., 2004) cells in this process. However, the precise molecular components and signaling pathways that are required to produce and maintain this microenvironment remain incompletely characterized.

In the hippocampus, precursor cells divide in the subgranular zone and migrate relatively short distances into the overlying granule cell layer, where they differentiate into new granule cell neurons (Cameron and Gould, 1994; Eriksson et al., 1998; Kornack and Rakic, 1999; Kuhn et al., 1996). In contrast, adult neuronal precursors destined for the olfactory bulb must migrate long distances first tangentially through the rostral migratory stream (RMS), and then radially into the outer layers of the olfactory bulb where they differentiate into two types of inhibitory interneurons: granule cells, and periglomerular cells (figure 1) (Lois and Alvarez-Buylla, 1994; Luskin, 1993). Adult generated neurons in the olfactory bulb appear to have distinct physiological characteristics, including increased sensitivity to activity-dependent regulation of surivival, compared to those generated by the early post-natal RMS (Corotto et al., 1994; Giachino et al., 2005; Lemasson et al., 2005; Winner et al., 2002). Work in our laboratory suggests that while precursors for both granule cells and periglomerular cells are located in the SVZ, there may separate pools of stem cells which give rise to each of these two neuronal populations (Beech et al., 2004). The transcription factor Pax6 appears to play a key role in specifying the facte of prospective dopamingeric periglomerular neurons (Hack et al., 2005; Kohwi et al., 2005). The mechanisms determining the fate of other types of newly generated neurons have not yet been determined.

Tangential migration in the adult RMS differs from radial migration of cortical neuronal precursors, which occurs during embryonic development, in that it does not require radial glial or axonal guides (Lois et al., 1996; Marin and Rubenstein, 2001; Marin and Rubenstein,

2003). However, migrating cells in the SVZ and RMS are ensheathed by a network of astrocytic cells (glial tubes) (Peretto et al., 1997). The cells comprising the glial tubes serve both as the precursors of, and a physical channel for, the migrating cells (Alvarez-Buylla and Garcia-Verdugo, 2002; Doetsch et al., 1999). Migration of neuronal precurosors through the RMS requires an ongoing rearrangement of the cytoskeleton (Ponti et al., 2004) as well as integration of chemo-attractive and repulsive cues from the environment to result in directed movement of the cell along the appropriate pathways.

Several recent papers have addressed the regulation of stem-cell proliferation in the SVZ by external factors (Hack et al., 2005; Liu et al., 2005; Lledo and Saghatelyan, 2005; Matarredona et al., 2005; Palma et al., 2005; Tonchev et al., 2005; Wang et al., 2005). The reader is referred to these papers for a more detailed discussion of factors involved in the regulation of proliferation in the SVZ. In this review I discuss what is known about the factors regulating the migration of adult neuronal precursors to the olfactory bulb, as well as recent unpublished data from our laboratory suggesting possible roles of specific membrane-cytoskeletal cross-linking proteins including members of the ERM (ezrin-radixin-moesin), cadherin, and catenin families in this process.

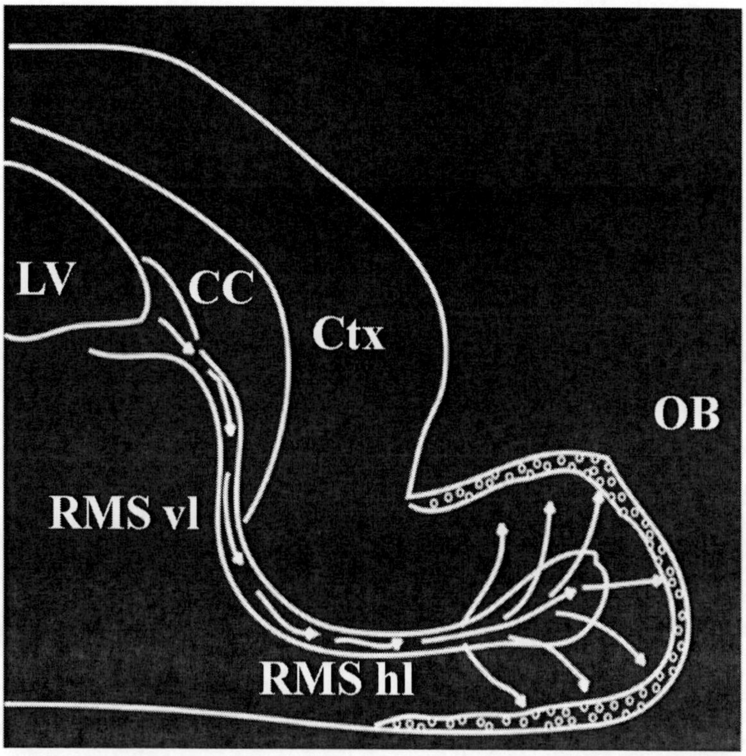

Figure 1. Schematic of the mouse forebrain illustrating the migratory pathway of SVZa derived neuroblasts en route to the olfactory bulb. Neuronal precursors destined for the olfactory bulb must migrate long distances first tangentially through the rostral migratory stream, and then radially into the outer layers of the olfactory bulb where they differentiate into two types of inhibitory interneurons: granule cells, and periglomerular cells . LV-Lateral Ventricle, CC- Corpus Callosum, CTx - Cerebral Cortex, RMS vl - Vertical Limb of Rostral Migratory Stream, RMS hl - Horizontal Limb of Rostral Migratory Stream, OB - Olfactory Bulb (adapted from (Coskun and Luskin, 2002)).

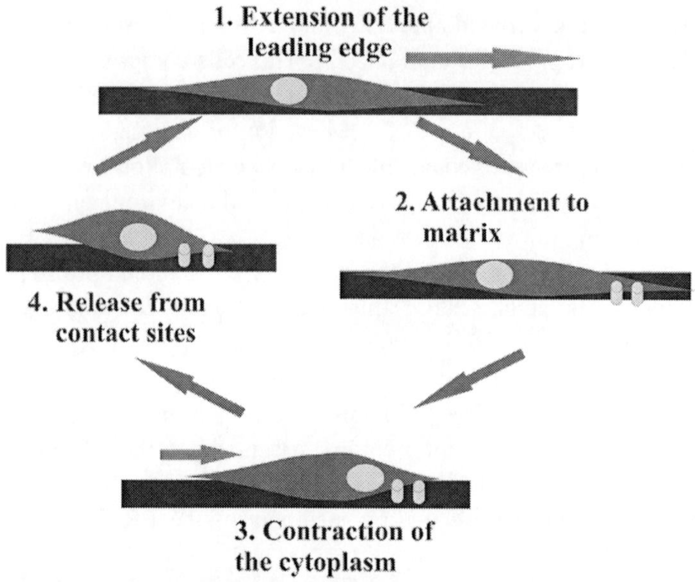

Figure 2. Cell migration as a cyclical process. Essential steps in the process of cell migration include: (1) extension of the leading edge; (2) adhesion to matrix contacts; (3) contraction of the cytoplasm; (4) release from contact sites and recycling of membrane receptors from the rear to the front of the cell. Each of these phases, in turn, is dependent upon multiple cyclical biochemical processes (see text) (adapted from (Sheetz et al., 1999)).

BASIC MECHANISMS OF CELL MIGRATION

Cell migration can be thought of as a cyclical process (Lauffenburger and Horwitz, 1996; Sheetz et al., 1999). Essential steps in this process include: (1) extension of the leading edge; (2) adhesion to matrix contacts; (3) contraction of the cytoplasm; (4) release from contact sites and recycling of membrane receptors from the rear to the front of the cell (figure 2). Each of these phases, in turn, is dependent upon multiple cyclical biochemical processes, many of which are regulated by members of the Rho family of GTPases (Raftopoulou and Hall, 2004; Ridley, 2001). Some of these pathways are described in more detail below.

1. Extension of the Leading Edge Is Regulated by Rho Family Members Rac1 and Cdc 42

Extension of the leading edge is driven by polymerization of actin filaments and bundling of these filaments into networks (Lodish et al., 2003) (figure 3). The growth of actin filaments occurs by the addition of ATP-actin subunits at the barbed, or (+) end of actin filaments, and the loss of subunits at the pointed, or (-) end. This process has been described as "actin treadmilling" (Pollard et al., 2000; Pollard and Borisy, 2003). Branching of filaments is mediated by activation of the Arp 2/3 complex, which binds to the sides or tip of existing filaments and induces the formation of a new daughter filament at a 70^0 angle from the mother filament (Welch and Mullins, 2002). These branches are stabilized by cortactin, and stabilization of the entire network occurs through cross-linking of the individual filaments by filamen A and α-actinin. Growth of fibers is terminated by the action of capping proteins, while members of the ADF/Cofilin family of proteins regenerate the pool of free actin monomers by severing filaments and promoting actin disassembly from the pointed (-) end. ATP-actin monomers are re-formed by the action of prolfilin, which promotes exchange of ADP for ATP.

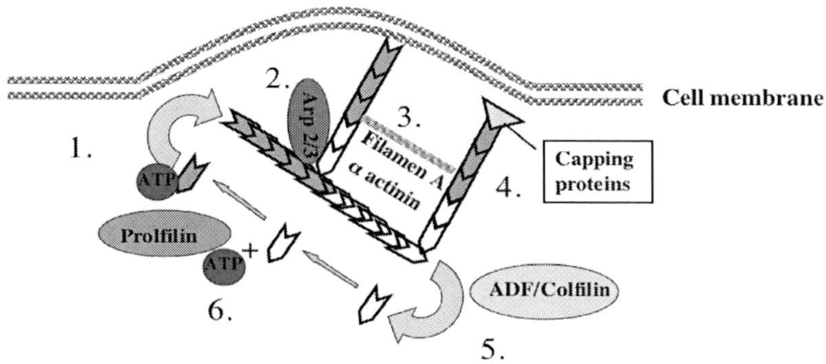

Figure 3. Extension of the leading edge is driven by actin treadmilling. Key steps in this process include: (1) Actin filaments grow by addition of ATP-actin to the barbed (+) end of the filament. (2) Branching of filaments is mediated by activation of the Arp 2/3 complex which binds to the sides or tip of existing filaments and induces the formation of a new daughter filament at a 70^0 angle from the mother filament. (3) These branches are stabilized by cross-linking of the individual filaments by filamen A and α-actinin. (4) Growth of fibers is terminated by the action of capping proteins. (5) Members of the ADF/Cofilin family of proteins regenerate the pool of free actin monomers by severing filaments and promoting actin disassembly from the pointed (-) end. (6) ATP-actin monomers are re-formed by the action of prolfilin, which promotes exchange of ADP for ATP (adapted from (Lodish et al., 2003)).

The rate of actin polymerization, and thus the rate of extension of the leading edge, is regulated by the Rho family of small GTP-binding proteins (Hall, 1998; Raftopoulou and Hall, 2004; Ridley, 2001; Ridley et al., 2003). The activity of Rho family GTPases, in turn is regulated by a number of different proteins including GEFs (Guanine-nucleotide exchange factors), GAPs (GTPase activating proteins) and GDIs (guanine-nucleotide dissociation inhibitors) (Symons and Settleman, 2000). These proteins allow Rho family members to cycle between active, GTP-bound forms, and inactive GDP-bound forms (figure 4A). Rho-family members, Rac1 and Cdc42 promote the assembly of actin at the leading edge (figure 4B). Activated forms of Rac1 and Cdc42 act through a number of effector proteins to increase the rate of actin polymerization and uncapping of actin filaments, while inhibiting actin depolymerization. Cdc42 activates N-WASP (Neural Wiskott-Aldrich syndrome protein). Rac1 activates IRSp53 (insulin receptor substrate p53), which forms a tri-molecular complex with the proteins N-WASP and WAVE (WASP family Verprolin-homologous protein), that activates the Arp 2/3 complex (Miki et al., 2000). Rac1 also activates PIP 5-kinase (phosphitidylinositol-4-phosphate 5-kinase), the enzyme that coverts PIP to PIP2 (phosphitidylinositol-4,5-bisphosphate), leading to barbed-end uncapping and severing of actin filaments (Hartwig et al., 1995). Finally, Cdc42 and Rac1 both activate PAK1 (p21-activated kinase) (Parrini et al., 2002). PAK1 activates LIMK (LIM domain kinase 1), which in turn, phosphorylates cofilin, inhibiting its actin-depolymerizing activity (Edwards et al., 1999; Maekawa et al., 1999).

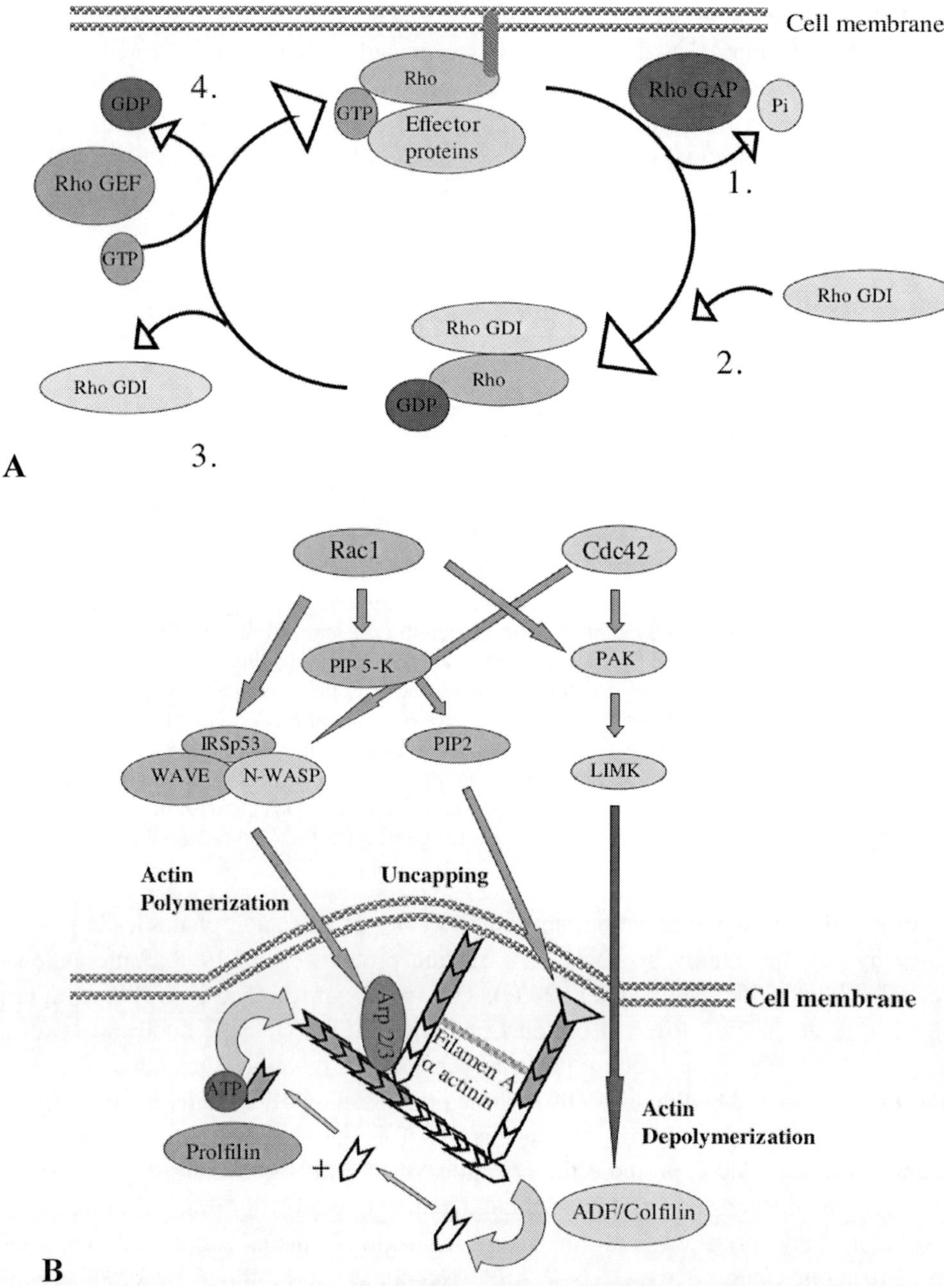

Figure 4. Role of Rho-family GTPases in actin assembly at the leading edge. (A) The activity of Rho-family GTPases is regulated by association with guanine-nucleotides. In their active state Rho family members are bound to GTP and form complexes with specific effector molecules (shown below in B). Binding of active Rho to specific membrane domains is mediated by a C-terminal prenyl group. Rho proteins become inactive following hydrolysis of GTP to GDP. (1) The intrinsic GTPase active of Rho proteins (and thus their rate of inactivation) is increased by proteins termed Rho GAPs (GTPase activating proteins). (2) Rho proteins can be maintained in an inactive, GDP-bound form when they form complexes with Rho GDIs (guanine-nucleotide dissociation inhibitors). (3) Dissociation of Rho GDIs allows Rho proteins to associate with GEFs (Guanine-nucleotide exchange factors), which return them (4) to the active, GTP-bound state by promoting the exchange of GDP for GTP (adapted from

(Watanabe et al., 2005)). (B) Rho-family members, Rac1 and Cdc42 promote the assembly of actin at the leading edge. Activated forms of Rac1 and Cdc42 act through a number of effector proteins to increase the rate of actin polymerization and uncapping of actin filaments, while inhibiting actin depolymerization. Effector proteins include IRSp53 (insulin receptor substrate p53), N-WASP (Neural Wiskott-Aldrich syndrome protein) and WAVE (WASP family Verprolin-homologous protein), which act on the Arp 2/3 complex to promote actin polymerization; PIP 5-kinase (phosphitidylinositol-4-phosphate 5-kinase), the enzyme that coverts PIP to PIP2 (phosphitidylinositol-4,5-bisphosphate), leading to barbed-end uncapping; and PAK1 (p21-activated kinase), which activates LIMK (LIM domain kinase 1), which in turn, phosphorylates cofilin , inhibiting its actin-depolymerizing activity (adapted from (Ridley, 2001)).

2. Integrins Regulate Cell Adhesion and Proliferation

Adhesion of cells to the extracellular matrix is mediated by a large family of heterodimeric receptors termed integrins. Integrin receptors consist of one each of 18 α and 8 β subunits, which assemble into at least 24 distinct integrins (Hynes, 2002). Integrins serve as mechanical links between the actin cytoskeleton and sites of extracellular contact, including cell-matrix and cell-cell contacts (reviewed in (Wiesner et al., 2005)). A number of important cellular functions including proliferation, differentiation, apoptosis, and cell migration are regulated by both integrins (Hynes, 2002) and the Rho family of GTPases (Hall, 1998; Ridley, 2001), and key members of the Rho GTPase family including Rho, Rac, and Cdc 42 are mediators of integrin signaling (Miranti and Brugge, 2002). Recent evidence has shown that integrins are essential in targeting Rho family GTPases to the plasma membrane via lipid rafts (del Pozo et al., 2004), as well as their coupling to downstream effectors. Activation of both Rac and Cdc42 requires dissociation of Rho-GDI (see figure 4A, above), and this dissociation is promoted by integrins (Del Pozo et al., 2002).

Binding of integrins to ECM proteins also leads, in a Rac-dependent manner, to the formation of focal complexes containing the membrane-cytoskeletal crosslinking proteins talin, vinculin, paxillin and phosphoproteins such as FAK (focal adhesion kinase) (Hynes, 2002; Zamir and Geiger, 2001). FAK binds to the cytoplasmic tails of β1 integrins, leading to autophosphorylation of FAK on Tyr^{397} and enhanced catalytic activity (Schaller, 2001). FAK binds and phosphorylates the p85 subunit of Pi3- Kinase (Phosphoinositide 3-OH kinase) (Chen and Guan, 1994) and the adapter protein Grb7 (growth factor receptor bound protein 7) (Shen and Guan, 2001). This phosphorylation pathway appears to promote cell migration, while an alternate pathway involving FAK mediated phosphorylation of the adapter protein Grb2 appears to promote cell-proliferation (Shen and Guan, 2001) (figure 5).

Different members of the β1 integrin family have been proposed to play specific roles in promoting migration or proliferation of neuronal stem cells (Jacques et al., 1998; Schmid and Anton, 2003; Tate et al., 2004), with α6β1 integrin acting to promote migration, while αvβ1 and α5β1 integrins were required for cell proliferation (Jacques et al., 1998). α1-, β8, and β1 subunits are expressed in the embryonic and early post -natal RMS, while αv and β6-integrins appear postnatally and persist into adulthood (Murase and Horwitz, 2002). α6β1 integrin is also expressed in the adult RMS, and appears to play a role in directional migration since injection of antibodies to α6β1 integrin disrupted neuronal migration, while injection of the α6β1 ligand, laminin, outside the RMS led to ectopic migration of cells towards the site of infusion (Emsley and Hagg, 2003). α3β1 integrins have also been proposed to play a role in the regulation of neuronal migration through their interaction with reelin and its downstream effector protein Dab1 (Dulabon et al., 2000; Forster et al., 2002; Schmid et al., 2005) (see section below on "Changing Directions").

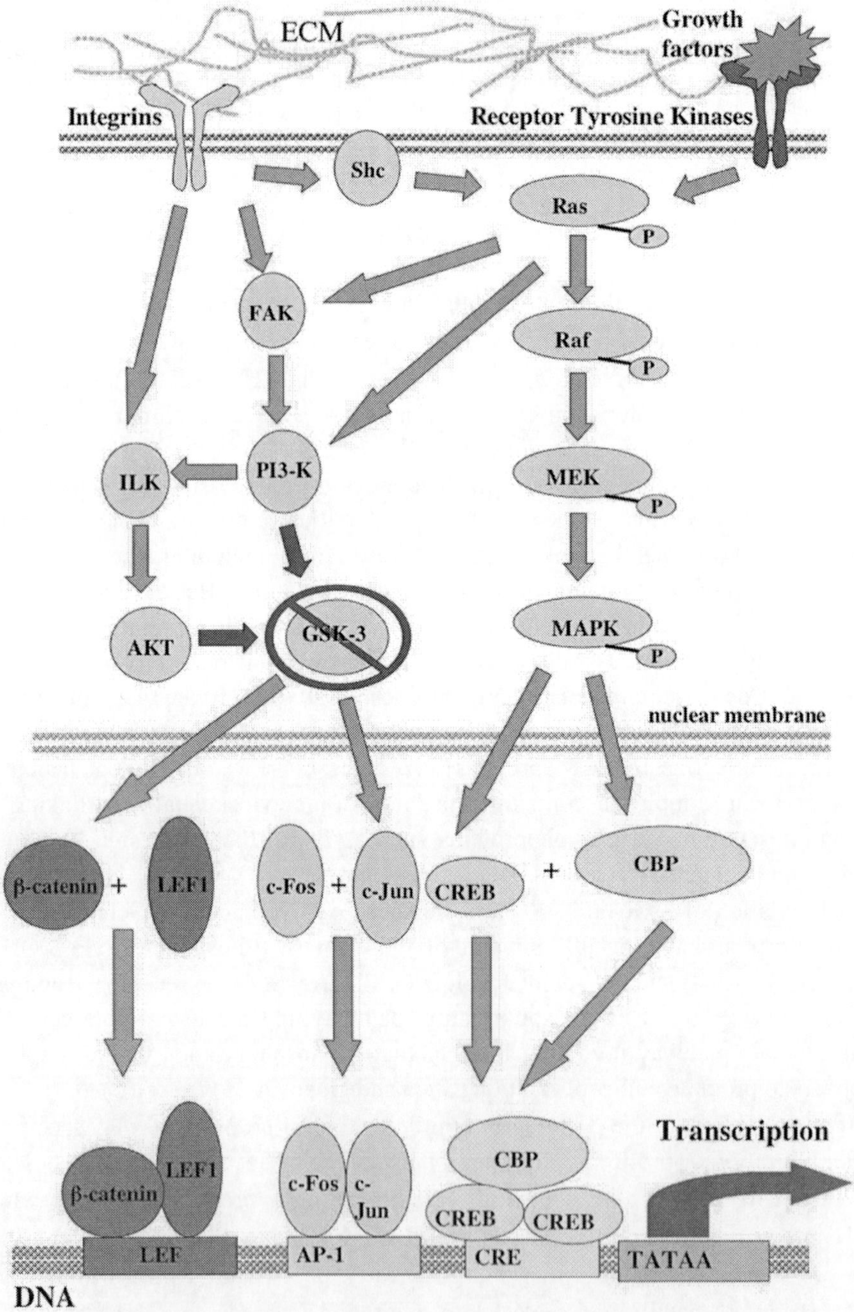

Figure 5. Cross talk between Integrin and Receptor Tyrosine Kinase pathways. Binding of integrins to ECM proteins leads, in a Rac-dependent manner to the formation focal complexes containing the membrane-cytoskeletal crosslinking proteins talin, vinculin, paxillin and phosphoproteins such as FAK (focal adhesion kinase). FAK binds to the cytoplasmic tails of β1 integrins, leading to

autophosphorylation on Tyr^{397} and enhanced catalytic activity (Schaller, 2001). FAK binds and phosphorylates the p85 subunit of Pi3- Kinase (Phosphoinositide 3-OH kinase)(Chen and Guan, 1994). This phosphorylation event appears to promote cell migration, while an alternate pathway involving FAK mediated phosphorylation of the adapter protein Grb2 (growth factor receptor bound protein 2) appears to promote cell-proliferation (Shen and Guan, 2001). Pi3-kinase catalyzes the formation of phosphatidylinositol (3,4,5) triphosphate (PIP3). PIP3 binds directly to ILK and stimulates the phosphorylation of AKT (protein kinase identified in the AKT virus, also known as protein kinase B) and GSK-3 (Glycogen synthase kinase 3) (Delcommenne et al., 1998). Activation of AKT activates a number of pathways inhibiting apoptosis and promoting cellular proliferation (Cantley, 2002; Coffer et al., 1998; Hemmings, 1997; Vanhaesebroeck and Alessi, 2000). Among them, AKT phosphorylates and inactivates GSK-3 (Cross et al., 1995), as well as the pro-apoptotic proteins BAD (bcl2 antagonist of cell death) (Downward, 1998)and caspace 9 (Cardone et al., 1998); cyclin-dependent kinase inhibitor, p27(Liang et al., 2002; Viglietto et al., 2002); and the forkhead transcription factor FKHRL-1 (Brunet et al., 1999) (adapted from (Dedhar, 2000)).

3. Contraction of the Cytoplasm Is Regulated by Actinomyosin Contractility and Rho

Actinomyosin contraction generates the force required for contraction of the cytoplasm and forward movement of the cell (Li et al., 2005). Actinomyosin contraction generates a traction force on the substrate at the sites of focal adhesions. This results in a counter-force that pulls the cell forward. Rho acts via ROCKs (Rho-kinases) to regulate actinomyosin contractility (Mitchison and Cramer, 1996; Rafelski and Theriot, 2004; Ridley et al., 2003). ROCKs both directly phosphorylate myosin light chains and phosphorylate and inhibit MLC phosphatase, leading to an increase in contractility (Amano et al., 2000; Kaibuchi et al., 1999). Myosin light chains are also phosphorylated by MLCK (myosin light chain kinase), which is activated by calcium-calmodulin (Stull et al., 1993) and inhibited by PAK (Sanders et al., 1999). MLCK can also be activated by phosphorylation by the MAPKs, ERK1 (extracellular signal regulated kinase) and ERK2, which results in increased MLC phosphorylation and enhanced cell migration (Cheresh et al., 1999; Klemke et al., 1997; Nguyen et al., 1999).

4. Cell Detachment and Recycling of Membrane Receptors Are Regulated by RhoA

Cell detachment is mediated in part by internalization of plasma membrane cholesterol and lipid raft associated proteins (del Pozo et al., 2004; Guan, 2004). While Rac1 and Cdc42 regulate the formation of focal adhesions at the front of migrating cells (see above), RhoA is required for the maturation of focal adhesions behind the leading edge and dissociation of focal adhesions at the rear (Kaverina et al., 2002).

Interactions between microtubules and actin also help to drive asymmetric microtubule assembly behavior in cells, such that microtubule growth occurs at the leading edge and microtubule shortening occurs at the cell body and rear (Waterman-Storer and Salmon, 1999). Microtubule growth and shortening, in turn, activate Rac1 and RhoA signaling, respectively, to control actin dynamics (Watanabe et al., 2005). Microtubule stabilization at the leading edge of migrating cells is regulated by Rho and its effector molecule Diaphanous (mDia) (Palazzo et al., 2001). Microtubule targeting of focal adhesions appears to promote the turnover of adhesions in protruding zones to allow further protrusion as well as the release of focal adhesions at retracting cell edges (Small and Kaverina, 2005).

Chain Migration in the RMS: Role of Homotypic Cell-Cell Interactions

The model of cell migration described above is based largely on studies of cultured fibroblasts migrating on a 2- dimensional surface containing one or more protein components of the extra-cellular matrix. Substrates that have been used in these studies include fibronectin, vitronectin, laminin, and collagen. However, it is important to note that cells migrating in a 3- dimensional matrix exhibit different properties form those migrating in 2 dimensions (Li et al., 2005). In particular, it has been suggested that the forces responsible for forward motion of migrating cells in the RMS result, not from cell-matrix, but from homotypic cell-cell interactions (Alvarez-Buylla, 1997; O'Rourke, 1996; Rousselot et al., 1995; Wichterle et al., 1997). These interactions appear to be mediated, at least in part, by the polysialylated form of neural cell adhesion molecule1 (PSA-NCAM) (O'Rourke, 1996). PSA-NCAM is expressed in the subventricular zone and along the length of the RMS (Bonfanti and Theodosis, 1994; Rousselot et al., 1995). Knockout mice, lacking either the entire NCAM1 gene (Cremer et al., 1994) or a specific 180 kilodalton isoform (Tomasiewicz et al., 1993), show abnormalities in neuronal migration, which result in a smaller olfactory bulb and an accumulation of neuronal precursors in the SVZ and proximal RMS. *In vitro* migration of SVZ derived cells can be blocked by enzymatic removal of the polysialic acid moiety from PSA-NCAM (Hu et al., 1996). Migrating cells in the RMS seem to have a similar requirement for PSA-NCAM *in vivo*, since removal of the PSA moieties from NCAM *in vivo* by injection of the enzyme endoneuraminidase N (endo N) into the intracranial cavity results in defects in migration that are similar to those seen in NCAM knockout mice (Ono et al., 1994). This suggests that the polysialic acid moiety is essential for neuronal migration in the RMS. Removal of PSA-NCAM also leads to premature maturation of SVZ-derived cells (Petridis et al., 2004). This effect is dependent on cell-cell contact, which suggests that one function of PSA-NCAM may be to prevent the premature activation of contact-dependent signaling cascades triggered by adhesion molecules.

Migrating cells in the RMS also express neural cell adhesion molecule L1 (L1CAM) (Jenkins et al., 2001). L1CAM is a member of the immunoglobulin superfamily of cell adhesion molecules (CAMs) that mediate cell-to-cell adhesion at the cell surface. L1CAMs mediate both homophilic (L1-L1) and heterophilic binding interactions (reviewed in (Haspel and Grumet, 2003)). L1CAM has been implicated in a large number of important functions in the development of the nervous system including guidance of neurite outgrowth in development, neuronal cell migration, axon bundling, synaptogenesis, myelination, neuronal cell survival, and long-term potentiation. Mutations in the L1CAM gene or deletion of the gene in knockout mice result in an X-linked mental retardation syndrome termed CRASH (corpus callosum agenesis, retardation, aphasia, spastic paraplegia, hydrocephalus)(reviewed in (Kenwrick et al., 2000)). L1CAM potentiates integrin-dependent cell migration on extracellular matrix proteins including fibronectin, vitronectin, and laminin, through a pathway involving c-Src, phosphatidylinositol-3 kinase, and MAP kinase (Thelen et al., 2002). Mutations in L1CAM associated with CRASH lead to loss of both L1-potentiated migration and MAP kinase activation, suggesting that defects in migration may contribute to some of the neurological deficits seen in these patients.

Surprisingly, in view of the fact that both PSA-NCAM and L1CAM both appear to promote cell migration, removal of PSA was found to augment L1-mediated adhesion (Acheson et al., 1991; Fujimoto et al., 2001). Thus, while both PSA-NCAM and L1CAM may contribute to homotypic interactions among migrating cells in the RMS, the nature of those interactions and the relationship of this type of migration to the substrate-dependent models described earlier remain incompletely understood.

Linking Directional Information to the Cytoskeleton

Migrating cells in the SVZ and RMS are guided by mechanisms similar to those used to guide developing axons to their appropriate targets. These mechanisms include chemo-attraction, chemo-repulsion, contact attraction, and contact repulsion (Guan and Rao, 2003; Park et al., 2002; Tessier-Lavigne and Goodman, 1996). In addition, activity-dependent mechanisms may play a role in determining the phenotype of newly generated prospective neurons, even while they are in the process of migrating and dividing along the course of the RMS (Bolteus and Bordey, 2004; Hack et al., 2005; Liu et al., 2005).

Factors which may play a role in directional migration in the RMS include netrins (Murase and Horwitz, 2002), semaphorins (Bagri and Tessier-Lavigne, 2002; He et al., 2002; Tamagnone and Comoglio, 2004), slits (Mason et al., 2001; Nguyen-Ba-Charvet et al., 2004; Wu et al., 1999), ephrins (Conover et al., 2000), and neuregulins (Anton et al., 2004). Netrins, slits and some semaphorins are secreted molecules that associate with cells or the extracellular matrix, while neuregulins, ephrins and other semaphorins are expressed on the cell surface (Falls, 2003; Yu and Bargmann, 2001). Each of these molecules interacts in various ways with members of the Rho GTPase family described in the first section. These interactions are thought to mediate the effects of the various chemoattractant and chemorepulsive factors of the actin cytoskeleton, and this in turn determines the effects of these factors on the cell's migration. Some of these interactions are described in more detail below.

Netrins

Netrins are large (approximately 70-80 kDa), soluble proteins that are closely related to the extracellular matrix molecule laminin (Tessier-Lavigne and Goodman, 1996; Yu and Bargmann, 2001). Netrins were originally characterized as chemo-attractants for spinal commissural axons during embryonic development (Serafini et al., 1996). However, subsequent work showed that the same netrin can be attractive to some axons and repulsive to others depending on which netrin receptors they express. Vertebrate receptors for netrins include Deleted in Colorectal Cancer (DCC) (Keino-Masu et al., 1996), neogenin (Yu and Bargmann, 2001), rostral cerebellar malformation (rcm) (Ackerman et al., 1997) and UNC-5H-1, -2, -3, and -4 (Engelkamp, 2002; Leonardo et al., 1997; Przyborski et al., 1998). Migrating cells in the adult RMS express both DCC and neogenin, and olfactory mitral cells (which are the synaptic targets of the newly generated granule and periglomerular cells in the bulb (Shepard and Greer, 1998)) express netrin-1 (Murase and Horwitz, 2002). *In vitro* experiments have shown that netrin-1 repels SVZa derived neuroblasts (Liu and Rao, 2003). However, interactions between netrins and DCC may have additional roles in regulating migration in the RMS, since addition of antibodies to DCC to slice cultures led migrating cells to move in a non-directed manner including back towards the SVZa or at right angles to the normal migratory pathway (Murase and Horwitz, 2002).

How is directional information from netrin-DCC interactions conveyed to the cytoskeleton? Netrin binding to DCC activates FAK and the tyrosine kinases Fyn and Src, stimulating tyrosine phosphorylation of these proteins as well as DCC itself (Li et al., 2004; Liu et al., 2004; Ren et al., 2004). Netrins also activate Rac and Cdc42 (Li et al., 2002; Shekarabi and Kennedy, 2002). As described above, the actions of these two proteins stimulate actin polymerization at the leading edge of migrating cells. While the majority of the work on signaling cascades downstream of netrin signaling has been done in the context of axonal guidance, it is likely that similar mechanisms underlie the requirement for DCC in directional movement in the RMS.

Semaphorins

Semaphorins are a family of highly conserved molecular signaling molecules that, like the netrins, were originally identified through their roles in axon guidance. Subsequent work showed that semaphorins are also involved in a number of other cellular functions including cell migration (reviewed in (Tamagnone and Comoglio, 2000; Tamagnone and Comoglio, 2004)). In contrast to netrins, semaphorins act primarily as chemo-repellants (Kolodkin et al., 1993; Pasterkamp and Kolodkin, 2003). Semaphorins include both membrane-associated and secreted molecules. All semaphorin family members have a secretion signal, a 500-amino acid sema domain, and 16 conserved cysteine residues. Semaphorin receptors include two families of proteins: the plexins and the neuropilins (Fujisawa and Kitsukawa, 1998). Membrane-bound semaphorins bind directly to plexins, while secreted semaphorins require neuropilins as co-receptors (Tamagnone and Comoglio, 2004). Semaphorin receptors can also form complexes with other transmembrane molecules including L1CAM (discussed above).

While there is currently no direct evidence for semaphorin-mediated signaling in directional guidance in the adult RMS, semaphorins are known to play key roles in the tangential migration of GABAergic interneurons during embryonic development (Marin et al., 2001; Tamamaki et al., 2003). Migrating interneurons expressing neuropilins are directed to the cortex, while those lacking them go to the striatum. This effect is mediated by a repulsive signal from the striatum, composed at least in part by Sema3A and Sema3F (Marin et al., 2001; Tamamaki et al., 2003). Sema3A is also expressed by mitral and tufted cells of the adult olfactory bulb (Giger et al., 1998) and numerous other members of the semaphorin family continue to be expressed in the adult brain (reviewed in (de Wit and Verhaagen, 2003)).

Many of signaling pathways implicated in cell migration can be regulated by semaphorins (Guan and Rao, 2003; Negishi et al., 2005; Tamagnone and Comoglio, 2004). Sema3A and Sema4D have been reported to inhibit both integrin-mediated cell adhesion and migration of endothelial cells (Barberis et al., 2004; Serini et al., 2003). Plexin activation in adherent cells leads to retraction of cellular processes and cell rounding in a process analogous to semaphorin-mediated "growth cone collapse" (Takahashi et al., 1999). This cellular collapse involves the disassembly of integrin-based focal adhesions as well as actin depolymerization.

Sema3A-induced growth cone collapse requires the activity of Rac1 (Jin and Strittmatter, 1997; Kuhn et al., 1999). The requirement for Rac1 activity in Sema3A-induced repulsive activity was a surprise since, as described above, Rac1 activity is usually associated with actin polymerization and extension of the leading edge. Rac1 is also required for Sema3A-induced endocytosis of growth cone plasma membranes and re-organization of F-actin during growth cone collapse, which may explain the requirement for Rac1 activity in Sema3A-mediated repulsion (Jurney et al., 2002).

Semaphorin signaling also involves activation of Rho. Plexin-B1 and Plexin-B2 activate Rho through two specific Rho-GEFs, PDZ-Rho-GEF and leukemia-associated Rho-GEF (LARG) (Perrot et al., 2002; Swiercz et al., 2002). Activation of these GEFs causes Rho to exchange GDP for GTP and leads to activation of Rho and Rho kinase. As described above, activation of Rho leads to cell detachment and recycling of membrane receptors, two processes that may be important in the repulsive effects of semaphorins on migrating cells.

Slit/Robo Proteins

Slits are secreted proteins, originally identified in Drosophila in a screen for genes involved in pattern formation, and subsequently shown to act as chemo-repellents for developing axons (Wong et al., 2002). Slits repel neuronal precursors migrating from the SVZa to the olfactory bulb (Wu et al., 1999), as well as precursors of GABAergic interneurons migrating from the lateral ganglionic eminence (LGE) to the neocortex (Zhu et al., 1999). Three slit genes have been cloned in mammals, Slit1, Slit 2, and Slit 3. Slit 1 is expressed primarily in brain, while Slit 2 is expressed primarily in the spinal cord and Slit3 is expressed in the thyroid gland (Itoh et al., 1998). Roundabouts (Robos) are cell surface receptors that are responsible for the repulsive effect Slit proteins in axon guidance (Brose et al., 1999; Long et al., 2004). Activation of Robo receptors by slit proteins inhibits N-cadherin mediated cell adhesion (Rhee et al., 2002) and can also silence netrin-mediated attractive signals (Stein and Tessier-Lavigne, 2001).

Slit1 and slit2 are expressed in the septum (Hu, 1999; Wu et al., 1999), while Robo2 and Robo3 are expressed by migrating cells in the RMS (Marillat et al., 2002). Both the septum and the choroid plexus exert endogenous chemo-repulsive activities on migrating SVZa derived neuroblasts (Hu, 1999; Wu et al., 1999). This activity consists of a combination of slit1 and slit2 (Nguyen-Ba-Charvet et al., 2004). A novel family of Rho GAPs, slit-robo GAPS (srGAPs), has been described that interacts selectively with the intracellular domain of Robo proteins (Wong et al., 2001). srGAP1 binds to and inhibits both Cdc42 and RhoA, but not Rac1. The effect of slits on migrating SVZa derived cells is lost in cells expressing a constitutively active form of Cdc42, indicating that down regulation of Cdc42 is crucial to the repulsive effects of slits on migrating neuroblasts (Wong et al., 2001). Surprisingly, slit1 is also expressed by the migrating neuroblasts themselves, as well transit amplifying cells, suggesting that the regulation of migration by slit-robo interactions is more complicated than simple repulsion from non-target tissues (Nguyen-Ba-Charvet et al., 2004).

Ephrins

Ephrins are ligands for the eph family of tyrosine kinases. Ephrins and ephs are involved in a number of important aspects of neural development including axon guidance, neural crest cell migration, and establishment of segmental boundaries in the developing nervous system. (Flanagan and Vanderhaeghen, 1998; O'Leary and Wilkinson, 1999; Palmer and Klein, 2003; Wilkinson, 2001). Ephrins and their receptors are divided into two classes based on their mode of attachment to the cell membrane and binding specificities. Ephrin-A ligands are anchored to the cell membrane by glycosylphosphatidylinositol (GPI) linkage, and bind to EphA receptors. Ephrin-B ligands are transmembrane proteins that bind EphB receptors. The EphA4 receptor binds to both classes of ephrin ligands. A striking feature of ephrin/eph signaling is the bi-directional nature of the signal. Ephrin binding induces dimerization of the eph receptor and tyrosine phosphorylation of the receptor itself as well as downstream substrates. In the case of ephrin-B/ephB binding, it also increases tyrosine phosphorylation of the intracellular domain of ephrin B. This phosphorylation is mediated by the Src family of tyrosine kinases (Palmer et al., 2002). Ephrin-A/ephA signaling also appears to be bi-directional and leads to an increase in integrin-mediated cell adhesion. This process requires the Fyn tyrosine kinase and a 120-kDa raft membrane protein that is tyrosine-phosphorylated specifically after ephrin-A activation (Davy et al., 1999; Davy and Robbins, 2000; Huai and Drescher, 2001).

Ephrin-B2 and ephrin-B3 are expressed by glial tube astrocytes in the SVZ and RMS, while EphA4, EphB2, and EphB3 are expressed by as yet unidentified cells in the RMS (Conover et al., 2000). Infusion of the extracellular portion of either EphB2 or ephrin-B2 into the lateral ventricles disrupts migration of neuronal precursors in the RMS and leads to an increase in the proliferation of SVZ astrocytes (Conover et al., 2000). As noted in the introduction, SVZ astrocytes serve both as the precursors of, and a physical channel for, the migrating cells (Alvarez-Buylla and Garcia-Verdugo, 2002; Doetsch et al., 1999). The fact that infusion of extracellular portions of either EphB2 or ephrin-B2 leads to a similar result suggests that bi-directional signaling between SVZ astrocytes and migrating neuroblasts may be important in regulating the balance between proliferation and migration of newly generated cells in the adult SVZ. The bi-directional nature of this signal may relate to the requirement for glial cells in maintaining an appropriate "niche" for adult neurogenesis (Lim and Alvarez-Buylla, 1999; Song et al., 2002).

Like the other proteins discussed above, eph/ephrin signaling is also coupled to cytoskeletal changes by the Rho family of GTPases. Eph/ephrin signaling activates a variety of exchange factors that regulate Rho family GTPases in neurons (reviewed in (Murai and Pasquale, 2005; Noren and Pasquale, 2004)). These exchange factors include ephexin1 (Sahin et al., 2005; Shamah et al., 2001), vav2 (Cowan et al., 2005), intersectin-1 (Irie and Yamaguchi, 2002), kalirin-7 (Penzes et al., 2003), and tiam1 (Tanaka et al., 2004). Ephexin1 is coupled to EphA receptors, while intersectin-1 and kalirin-7 are coupled to EphB, and tiam1 and vav2 can be coupled to either EphA or EphB (Murai and Pasquale, 2005).

The activation of EphA receptors induces the activation of RhoA and the inhibition of Rac1 (Wahl et al., 2000). These effects are mediated by ephexin1, a GEF for Rho that interacts directly with the kinase domain of EphA receptors. Stimulation of EphA receptors by Ephrin-A leads to tyrosine phosphorylation of ephexin1 and increases the GEF the activity of ephexin1 leading to RhoA activation (Sahin et al., 2005). Other results suggest that Rac1 is required for EphA mediated effects on endocytosis of the growth cone plasma membrane and reorganization of F-actin (Jurney et al., 2002). Thus, both RhoA and Rac1 may be required for ephrin-mediated effects on cell migration.

Vav exchange factors are also known to activate Rho GTPases (Schmidt and Hall, 2002), and tiam1 has been shown to activate Rac1 downstream of ephrinB1 (Tanaka et al., 2004). As noted above, Rac1 is thought to play an important role in regulating extension of the leading edge in migrating cells, while Rho plays important roles in regulating both cell contraction and cell detachment. However, roles for specific GEFs have not yet been identified in relation to the effects of EphrinB2 and EphB2 on cell migration and proliferation in the adult SVZ and RMS.

Neuregulins

The neuregulins (NRGs) are cell-cell signaling proteins that are ligands for receptor tyrosine kinases of the ErbB family (reviewed in (Buonanno and Fischbach, 2001; Carpenter, 2003; Falls, 2003; Riese and Stern, 1998; Yarden and Sliwkowski, 2001)). ErbB4 is expressed in migrating neuroblasts in the SVZ and RMS of adult mice (Anton et al., 2004), as well as other populations of neuronal precursors that undergo tangential migration (Flames et al., 2004; Yau et al., 2003). ErbB4 can be activated by a large number of ligands including neuregulins (NRG1- NRG4), betacellulin, HB-EGF, and epiregulin. Following ligand binding, the ErbB4-ectodomain is cleaved by the metalloprotease TACE (tumor necrosis factor-alpha converting enzyme) (Rio et al., 2000; Zhou and Carpenter, 2000). A subsequent cleavage by gamma-secretase releases the ErbB4 intracellular domain from the membrane and facilitates its translocation to the nucleus, where it may influence gene transcription (Carpenter, 2003; Ni et al., 2001).

NRG1 is expressed in the SVZ and RMS, and *in vitro* experiments showed that NRG1 acts as a preferred migration substrate for neuroblasts derived from normal mice, but not those derived from mice with inducible mutations in the ErbB4 gene (Anton et al., 2004). Genetic linkage studies have implicated the NRG1 gene as a candidate gene for risk for schizophrenia (Arnold et al., 2005; Stefansson et al., 2003; Stefansson et al., 2002; Zhao et al., 2004), and mutant mice heterozygous for either NRG1 or its receptor, ErbB4, show a behavioral phenotype that overlaps with mouse models for schizophrenia (Stefansson et al., 2002). ErbB4 mutant mice also showed deficits in chain migration characterized by a shift from chain migration to migration as single cells, and a slower rate of migration and loss of directionality in slice culture. NRG1 also as acts as an attractant for GABAergic cortical interneurons undergoing tangential migration (Flames et al., 2004). These observations suggest that abnormal neuronal migration that results from defects in NRG1/ErbB4 signaling may contribute to the risk of developing schizophrenia.

NRG1 stimulation has been shown to induce the formation of a multiprotein complex that contains the tyrosine kinase Pyk2 as well as p190 RhoGAP (p190), RasGAP, c-Abl and the focal adhesion molecules p130cas and paxillin (Zrihan-Licht et al., 2004). This is consistent with previous work showing that NRG1 can regulate cytoskeletal reorganization and cell migration through PAK1 and Pi3-K (Adam et al., 1998). As discussed above, Pi3-K is downstream of FAK and paxillin in the integrin signaling pathway, and PAK1 can activate LIMK, which in turn, phosphorylates cofilin, inhibiting its actin-depolymerizing activity and thus promoting cell migration (Edwards et al., 1999; Maekawa et al., 1999).

Changing Directions: The Transition from Tangential to Radial Migration

The activities of the various signaling molecules discussed above including netrins, semaphorins, slits, ephrins, and neuregulins, all relate to the tangential migration of cells from the SVZ to the olfactory bulb along the RMS. However, before integrating into the synaptic architecture of the bulb, the migrating cells must leave the RMS and migrate to the appropriate cell layers in the olfactory bulb where they then differentiate into two types of interneurons: granule cells and periglomerular cells (Lois and Alvarez-Buylla, 1994; Luskin, 1993). Two molecules in the extracellular matrix, reelin (Hack et al., 2002) and Tenascin-R (Saghatelyan et al., 2004), as well as the secreted factor prokineticin-2 (Ng et al., 2005) have been proposed to key roles in the transition from radial to tangential migration. Below, I discuss what is known about each of these molecules as it relates to the transition from tangential to radial migration in the adult RMS.

Reelin is a large extra-cellular matrix protein that is important in the formation of normal layered structures in the brain during embryonic development (Lambert de Rouvroit and Goffinet, 1998; Rugarli and Ballabio, 1995). Mice lacking both copies of the reelin gene develop "inside-out" brains, with a reversal of the normal layers of the cerebral cortex (Caviness and Sidman, 1972; Rakic and Caviness, 1995; Rice and Curran, 2001). Reelin mRNA (Alcantara et al., 1998) and protein (Pappas et al., 2003) are expressed in the adult olfactory bulb, and reeler mice exhibit abnormalities in the architecture of the olfactory bulb similar to those seen in the cerebral cortex (Caviness and Sidman, 1972). The reelin protein has been proposed to serve as a detachment signal during the migration of neuronal precursors from the subventricular zone to the olfactory bulb in the adult brain (Hack et al., 2002). Immuno-labeling for the reelin protein in heterozygous mice shows an accumulation of reelin-positive cells in the rostral migratory stream that are apparently unable to shift from tangential to radial migration (Pappas et al., 2003). Reeler heterozygous mice also exhibit deficits in olfactory learning (Larson et al., 2003), suggesting that this abnormal migration has functional consequences in the adult brain.

Reelin receptors include the very low density lipoprotein receptor (VLDLR) and the apoE receptor 2 (ApoER2) (Hiesberger et al., 1999). Binding of reelin to these receptors induces tyrosine phosphorylation of the cytosolic adaptor protein, Disabled-1 (Dab1). "Scrambler" (Ware et al., 1997) or "yotari" (Sheldon et al., 1997) mice have mutations in the Dab1 gene and have a phenotype identical to that of the reeler mouse, demonstrating that the tyrosine phosphorylation of Dab1 mediates the biological effects of reelin signaling. *In vitro* experiments in a Matrigel culture system (Hack et al., 2002) showed that reelin acted as a detachment signal for chain-migrating interneurons precursors in the olfactory bulb, inducing their dispersal into individual cells.

As discussed above, reelin has been shown to associate with α3β1 integrins. Experiments using α3β1 integrin-deficient mouse embryos showed that effects of reelin on cortical neuronal migration were dependent on interactions between reelin and α3β1 integrins (Dulabon et al., 2000). A similar requirement for reelin-integrin interactions has been shown in the formation of the radial glial scaffold in the hippocampus (Forster et al., 2002). These authors also showed that recombinant Reelin induced branching of glial processes in vitro, suggesting that reelin may play a role in regulating interactions between migrating neurons and radial glia. Subsequent work has shown that α3β1 integrins binds to the N-terminal region of reelin, a site distinct from the region of reelin shown to associate with VLDLR and ApoER2, while Dab1binds to the cytoplasmic region of β1 integrins in a reelin-dependent manner (Schmid et al., 2005). These observations suggest that α3β1 integrins mediate the effects of reelin in the transition from tangential to radial migration independently of other reelin receptors such as VLDLR and ApoER2.

Tenascin-R (TNR) is a member of the tenascin gene family (reviewed in (Jones and Jones, 2000; Pesheva and Probstmeier, 2000)). TNR isoforms are generated by pre-mRNA alternative splicing. In the adult mouse TNR immunostaining is restricted to the granule cells and internal plexiform layers of the olfactory bulb (Pesheva and Probstmeier, 2000). Expression of TNR is activity dependent and is markedly reduced by odor deprivation (Saghatelyan et al., 2004). Grafting of TNR-transfected cells into non-neurogenic regions rerouted migrating neuroblasts toward these regions, while in TNR deficient mice neuroblasts were retained in the anterior segment of the rostral migratory stream (Saghatelyan et al., 2004). These observations suggest that tenascin-R may mediate both the detachment of migrating neuroblasts from chains and the initiation of radial migration.

Experiments in other contexts (reviewed in (Pesheva and Probstmeier, 2000)) show that TNR can either enhance or inhibit cell-matrix and neuronal-glial interactions. These effects are both substrate-dependent and isoform-specific. The specific molecules linking TNR to cytoskeletal changes in migrating cells have not yet been determined, but they may involve effects on β1 integrins and downstream phosphorylation of FAK since in other contexts TNR can inhibit both β1 integrin-dependent cell adhesion and tyrosine phosphorylation of FAK (Pesheva and Probstmeier, 2000).

Prokineticin-2 (PK2) is a cysteine rich secreted protein that, along with the related protein prokineticin-1, has been implicated a wide variety of biological functions including contraction of gastrointestinal smooth muscle (Li et al., 2001), regulation of circadian rhythms (Cheng et al., 2002), and regulation of proliferation, survival and migration of endothelial cells (LeCouter et al., 2003). Prokineticins are thought to exert their biological effects by activation of two highly related G-protein coupled receptors (Soga et al., 2002). Activation of prokineticin receptors leads to mobilization of calcium, stimulation of phosphoinositide turnover, and activation of p44/p42 MAPK signaling pathways (Lin et al., 2002).

PK2 is expressed in both periglomerular and granule cells, the two types of adult generated neurons in the olfactory bulb, while its receptors, PKR1 and PKR2, are expressed in the underlying RMS (Ng et al., 2005). PK2 knockout mice have smaller olfactory bulbs, and neuroblasts in these animals fail to make the transition from tangential to radial migration. *In vitro* PKs act as chemoattractants for SVZa derived neuroblasts, and also lead to the dispersal of migrating chains of neuroblasts (Ng et al., 2005). The mechanisms linking PK2 signaling to changes in cytoskeletal dynamics are currently unknown, as are the relationships among PK2, reelin and TNR, all of which have been implicated in the transition from tangential to radial migration in the adult RMS.

Balancing Proliferation and Migration

Unlike the neuronal precursors that give rise to cortical neurons during embryonic development, which divide in the ventricular zone and then migrate to appropriate positions in the overlying cortex, neuronal precursors in the RMS continue to divide as they migrate (Menezes et al., 1995; Rousselot et al., 1995). Maintaining an appropriate balance between proliferation and migration necessitates a series of signaling pathways which co-ordinate these activities. Experiments in a variety of systems have shown that the processes of migration and cell-cycle exit can be coupled (Choi et al., 2005; Conover et al., 2000; Liu et al., 2005). As discussed above, infusion of the extracellular portion of either EphB2 or ephrin-B2 into the lateral ventricles both disrupts migration of neuronal precursors in the RMS and leads to an increase in the proliferation of SVZ astrocytes (Conover et al., 2000). This suggests that bi-directional signaling between SVZ astrocytes and migrating neuroblasts is important in maintaining the balance between migration and proliferation.

Recent work has shown that both the speed of migration of SVZa derived neuroblasts (Bolteus and Bordey, 2004), and the proliferation of SVZ astrocytes (Liu et al., 2005) is regulated by non-synaptic GABA released by migrating neuroblasts. SVZ progenitor cells express the GABA transporter (GAT) (Wang et al., 2003), and uptake of GABA by this transporter is thought to increase the speed of migration by relieving the inhibitory effects of extra-synaptic GABA on the migrating neuroblasts (Bolteus and Bordey, 2004). In addition, depolarization of SVZ precursors leads to a decrease in the rate of proliferation (Liu et al., 2005). These results underscore the importance of bi-directional signaling mechanisms in regulating both proliferation and migration in the SVZ and RMS.

Recent data from our laboratory (Cleary et al., unpublished data) indicate that SVZ astrocytes express two proteins, ezrin and beta-catenin, that may be important in this process. Ezrin is a member of the highly conserved ERM (Ezrin-Radixin-Moesin), band 4.1 family of membrane-cytoskeletal linking proteins (Gould et al., 1989; Takeuchi et al., 1994). ERM proteins are involved in a wide variety of cellular functions including cell motility, signal transduction, cell-cell and cell-matrix recognition (Bretscher et al., 1997) (Bretscher et al., 2002; Louvet-Vallee, 2000). Ezrin is expressed in radial glia (Johnson et al., 2002), and co-localizes with L1CAM in migrating cortical neurons (Mintz et al., 2003). Ezrin binds the intracellular domain of L1CAM (Dickson et al., 2002). This interaction is required for L1CAM-mediated homotypic interactions (Cheng et al., 2005a). As discussed L1CAM-mediated homotypic interactions may play a role in regulating migration in the RMS. Ezrin also binds to Rho-GDI (Hirao et al., 1996; Maeda et al., 1999; Takahashi et al., 1997). Binding of Ezrin to Rho-GDI allows Rho to dissociate from Rho-GDI and become active. In a positive feedback loop, ezrin activity is regulated by phosphorylation by Rho-kinase (reviewed in (Bretscher et al., 2000; Bretscher et al., 2002; Gautreau et al., 2002)).

β-Catenin, in association with the transcription factor LEF1 (Porfiri et al., 1997), is a key downstream mediator of the wnt signaling pathway (Logan and Nusse, 2004). Wnt signaling inhibits neural differentiation of embryonic stem cells by up-regulating the expression bone morphogenetic proteins (BMPs) (Haegele et al., 2003). Endogenous β-catenin is highly expressed in embryonic neural precursors, and transgenic mice that express high levels of a stabilized β-catenin transgene develop enlarged brains with expanded precursor populations (Chenn and Walsh, 2002). Adult mice expressing the same β-catenin transgene develop enlarged forebrains with expanded subventricular zones (Chenn and Walsh, 2003). These results suggest that nuclear β-catenin can promote cell proliferation and inhibit differentiation of stem cells in the adult SVZ. In other cell types ezrin forms a complex with E-cadherin and β-catenin that appears to regulate both cell-cell and cell-matrix recognition (Hiscox and Jiang, 1999). Disruption of this complex by tyrosine phosphorylation of ezrin leads to loss of cell-cell adhesion and increased invasiveness. Decreased expression of E-cadherin in breast cancer cells is also associated with an increase in nuclear β-catenin, a loss of cell-cell adhesion, and an increased rate of cell division (Yang et al., 2001). These observations suggest that sequestration of β-catenin in membrane-associated complexes can serve to regulate the rate of cell division.

Cadherins are calcium-dependent homophilic adhesion molecules that, together with their intracellular binding partners the catenins, are thought to mediate cell-cell adhesion and migration (reviewed in (Nelson and Nusse, 2004; Redies, 2000)). Unlike most cadherins, which show highly restricted patterns of expression during development, N-cadherin is ubiquitously expressed in the entire proliferative neuroepithelium. N-cadherin and β-catenin are co-localized in axons and dendrites prior to synapse formation (Benson and Tanaka, 1998). A complex containing N-cadherin and β-catenin has also been suggested to play a role in down-regulating cell proliferation and inducing terminal differentiation in myoblasts (Gavard et al., 2004). The fact that over-expressing β-catenin leads to increased proliferation in the adult SVZ (Chenn and Walsh, 2003) suggests that the ratio of nuclear to membrane-associated β-catenin may be important in balancing ongoing proliferation and migration in the adult SVZ and RMS as well.

Migrating cells in the RMS also express p120 catenin (Chauvet et al., 2003). Like β-catenin, p120 catenin can bind to the intracellular portion of both E-cadherin and N-cadherin, as well as to β-catenin (Chauvet et al., 2003; Daniel and Reynolds, 1995; Reynolds et al., 1994; Shibamoto et al., 1995; Staddon et al., 1995). p120 catenin has also been shown to inhibit RhoA (Anastasiadis et al., 2000). Moreover, inhibition of RhoA and binding to cadherins was mutually exclusive, which would allow RhoA to become activated specifically at nascent cell-cell contacts. As discussed above, RhoA is involved in multiple aspects of cell migration including regulation of actinomyosin contractility as well as cell detachment and recycling of membrane receptors. Disruption of stress fibers and focal adhesions by p120 catenin can be blocked by dominant negative Cdc42 and Rac1 or by constitutively active Rho-kinase, but is enhanced by dominant negative RhoA (Grosheva et al., 2001; Noren et al., 2000). These effects are mediated by binding and activation of the RhoGEF, vav2 (Noren et al., 2000) (discussed above, in relation to eph/ephrin signaling). Co-transfection with an excess of E-cadherin leads to sequestration of exogenous p120ctn to cell-cell junctions or to small cadherin-containing vesicles, and abolishes p120ctn effects on cell morphology (Grosheva et al., 2001). Thus, interactions among cadherins and catenins, and perhaps other cytoskeletal crosslinking proteins that effect the subcellular localization of these proteins, appear to be mechanisms by which both proliferation and migration in the SVZ and RMS may be regulated.

CONCLUSION

Neuronal precursors of adult generated olfactory bulb interneurons migrate long distances first tangentially through the rostral migratory stream, and then radially into the outer layers of the olfactory bulb where they differentiate into two types of inhibitory interneurons: granule cells, and periglomerular cells. This migration requires an ongoing rearrangement of the cytoskeleton as well as integration of chemo-attractive and repulsive cues from the environment to result in directed movement of the cell along the appropriate pathways. Cell migration can be thought of as a cyclical process. Essential steps in this process include: (1) extension of the leading edge; (2) adhesion to matrix contacts; (3) contraction of the cytoplasm; (4) release from contact sites and recycling of membrane receptors from the rear to the front of the cell. Each of these phases, in turn, is dependent upon multiple cyclical biochemical processes, many of which are regulated by members of the Rho family of GTPases. Directional cues present in the environment of the migrating cells exert many of their effects via these same GTPases. Netrins, semaphorins, slits, ephrins, and neuregulins are among the molecules proposed to regulate the directional movement of SVZa derived neuroblasts along the RMS. Each of these families of signaling molecules has been shown to interact with one or more of the Rho family GTPases to bring about their effects on cell migration.

Two molecules in the extracellular matrix, reelin and TNR, and the secreted protein, PK2, have been shown to key roles in the transition from radial to tangential migration. Reelin is thought to exert its effects through interactions with $\alpha3\beta1$ integrins. The mechanism by which TNR exerts its effects is unknown, but may involve interactions with $\beta1$ integrins as well. PK2 signals through two G-coupled receptors that activate the MAPK pathway. However, the mechanisms linking PK2 to cytoskeletal changes have not yet been described.

A central theme of the work described here is that signaling molecules that regulate both cell proliferation and migration do so, in large part, via their effects on the cytoskeleton. Mechanisms involved in regulating the balance between proliferation and migration in include bi-directional signaling between SVZ astrocytes and migrating neuroblasts mediated by ephs and ephrins, and GABAergic signaling from neuroblasts to astrocytes. Membrane-cytoskeletal crosslinking proteins including members of the ERM, catenin and cadherin families, may also play important roles in balancing proliferation and migration of newly generated cells in the adult SVZ and RMS. Understanding the links between specific signaling events and the structural changes they induce in stem cells and their progeny represents one of the major challenges for the next decade. This understanding will help us as we attempt to harness the restorative capacity inherent in stem cells for therapeutic purposes in the CNS and elsewhere in the body.

REFERENCES

Acheson, A., Sunshine, J. L., and Rutishauser, U. (1991). NCAM polysialic acid can regulate both cell-cell and cell-substrate interactions. *J Cell Biol 114*, 143-153.

Ackerman, S. L., Kozak, L. P., Przyborski, S. A., Rund, L. A., Boyer, B. B., and Knowles, B. B. (1997). The mouse rostral cerebellar malformation gene encodes an UNC-5-like protein. *Nature 386*, 838-842.

Adam, L., Vadlamudi, R., Kondapaka, S. B., Chernoff, J., Mendelsohn, J., and Kumar, R. (1998). Heregulin regulates cytoskeletal reorganization and cell migration through the p21-activated kinase-1 via phosphatidylinositol-3 kinase. *J Biol Chem 273*, 28238-28246.

Alcantara, S., Ruiz, M., D'Arcangelo, G., Ezan, F., de Lecea, L., Curran, T., Sotelo, C., and Soriano, E. (1998). Regional and cellular patterns of reelin mRNA expression in the forebrain of the developing and adult mouse. *J Neurosci 18*, 7779-7799.

Alvarez-Buylla, A. (1997). Mechanism of migration of olfactory bulb interneurons. *Semin Cell Dev Biol 8,* 207-213.

Alvarez-Buylla, A., and Garcia-Verdugo, J. M. (2002). Neurogenesis in adult subventricular zone. *J Neurosci 22*, 629-634.

Amano, M., Fukata, Y., and Kaibuchi, K. (2000). Regulation and functions of Rho-associated kinase. *Exp Cell Res 261*, 44-51.

Anastasiadis, P. Z., Moon, S. Y., Thoreson, M. A., Mariner, D. J., Crawford, H. C., Zheng, Y., and Reynolds, A. B. (2000). Inhibition of RhoA by p120 catenin. *Nat Cell Biol 2*, 637-644.

Anton, E. S., Ghashghaei, H. T., Weber, J. L., McCann, C., Fischer, T. M., Cheung, I. D., Gassmann, M., Messing, A., Klein, R., Schwab, M. H., et al. (2004). Receptor tyrosine kinase ErbB4 modulates neuroblast migration and placement in the adult forebrain. *Nat Neurosci 7*, 1319-1328.

Arnold, S. E., Talbot, K., and Hahn, C. G. (2005). Neurodevelopment, neuroplasticity, and new genes for schizophrenia. *Prog Brain Res 147*, 319-345.

Bagri, A., and Tessier-Lavigne, M. (2002). Neuropilins as Semaphorin receptors: in vivo functions in neuronal cell migration and axon guidance. *Adv Exp Med Biol 515*, 13-31.

Barberis, D., Artigiani, S., Casazza, A., Corso, S., Giordano, S., Love, C. A., Jones, E. Y., Comoglio, P. M., and Tamagnone, L. (2004). Plexin signaling hampers integrin-based adhesion, leading to Rho-kinase independent cell rounding, and inhibiting lamellipodia extension and cell motility. *FASEB J 18*, 592-594.

Beech, R. D., Cleary, M. A., Treloar, H. B., Eisch, A. J., Harrist, A. V., Zhong, W., Greer, C. A., Duman, R. S., and Picciotto, M. R. (2004). Nestin promoter/enhancer directs transgene expression to precursors of adult generated periglomerular neurons. *J Comp Neurol 475*, 128-141.

Benson, D. L., and Tanaka, H. (1998). N-cadherin redistribution during synaptogenesis in hippocampal neurons. *J Neurosci 18*, 6892-6904.

Bolteus, A. J., and Bordey, A. (2004). GABA release and uptake regulate neuronal precursor migration in the postnatal subventricular zone. *J Neurosci 24*, 7623-7631.

Bonfanti, L., and Theodosis, D. T. (1994). Expression of polysialylated neural cell adhesion molecule by proliferating cells in the subependymal layer of the adult rat, in its rostral extension and in the olfactory bulb. *Neuroscience 62*, 291-305.

Bretscher, A., Chambers, D., Nguyen, R., and Reczek, D. (2000). ERM-Merlin and EBP50 protein families in plasma membrane organization and function. *Annu Rev Cell Dev Biol 16*, 113-143.

Bretscher, A., Edwards, K., and Fehon, R. G. (2002). ERM proteins and merlin: integrators at the cell cortex. *Nat Rev Mol Cell Biol 3*, 586-599.

Bretscher, A., Reczek, D., and Berryman, M. (1997). Ezrin: a protein requiring conformational activation to link microfilaments to the plasma membrane in the assembly of cell surface structures. *Journal of Cell Science 110*, 3011-3018.

Brose, K., Bland, K. S., Wang, K. H., Arnott, D., Henzel, W., Goodman, C. S., Tessier-Lavigne, M., and Kidd, T. (1999). Slit proteins bind Robo receptors and have an evolutionarily conserved role in repulsive axon guidance. *Cell 96*, 795-806.

Brunet, A., Bonni, A., Zigmond, M. J., Lin, M. Z., Juo, P., Hu, L. S., Anderson, M. J., Arden, K. C., Blenis, J., and Greenberg, M. E. (1999). Akt promotes cell survival by phosphorylating and inhibiting a Forkhead transcription factor. *Cell 96*, 857-868.

Buonanno, A., and Fischbach, G. D. (2001). Neuregulin and ErbB receptor signaling pathways in the nervous system. *Curr Opin Neurobiol 11*, 287-296.

Cameron, H. A., and Gould, E. (1994). Adult neurogenesis is regulated by adrenal steroids in the dentate gyrus. *Neuroscience 61*, 203-209.

Cantley, L. C. (2002). The phosphoinositide 3-kinase pathway. *Science 296*, 1655-1657.

Cardone, M. H., Roy, N., Stennicke, H. R., Salvesen, G. S., Franke, T. F., Stanbridge, E., Frisch, S., and Reed, J. C. (1998). Regulation of cell death protease caspase-9 by phosphorylation. *Science 282*, 1318-1321.

Carpenter, G. (2003). ErbB-4: mechanism of action and biology. *Exp Cell Res 284*, 66-77.

Caviness, V. S., Jr., and Sidman, R. L. (1972). Olfactory structures of the forebrain in the reeler mutant mouse. *J Comp Neurol 145*, 85-104.

Chauvet, N., Prieto, M., Fabre, C., Noren, N. K., and Privat, A. (2003). Distribution of p120 catenin during rat brain development: potential role in regulation of cadherin-mediated adhesion and actin cytoskeleton organization. *Mol Cell Neurosci 22*, 467-486.

Chen, H. C., and Guan, J. L. (1994). Association of focal adhesion kinase with its potential substrate phosphatidylinositol 3-kinase. *Proc Natl Acad Sci U S A 91*, 10148-10152.

Cheng, L., Itoh, K., and Lemmon, V. (2005a). L1-Mediated Branching Is Regulated by Two Ezrin-Radixin-Moesin (ERM)-Binding Sites, the RSLE Region and a Novel Juxtamembrane ERM-Binding Region. *J Neurosci 25*, 395-403.

Cheng, L. C., Tavazoie, M., and Doetsch, F. (2005b). Stem cells from epigenetics to microRNAs. *Neuron 46*, 363-367.

Cheng, M. Y., Bullock, C. M., Li, C., Lee, A. G., Bermak, J. C., Belluzzi, J., Weaver, D. R., Leslie, F. M., and Zhou, Q. Y. (2002). Prokineticin 2 transmits the behavioural circadian rhythm of the suprachiasmatic nucleus. *Nature 417*, 405-410.

Chenn, A., and Walsh, C. A. (2002). Regulation of cerebral cortical size by control of cell cycle exit in neural precursors. *Science 297*, 365-369.

Chenn, A., and Walsh, C. A. (2003). Increased neuronal production, enlarged forebrains and cytoarchitectural distortions in beta-catenin overexpressing transgenic mice. *Cereb Cortex 13*, 599-606.

Cheresh, D. A., Leng, J., and Klemke, R. L. (1999). Regulation of cell contraction and membrane ruffling by distinct signals in migratory cells. *J Cell Biol 146*, 1107-1116.

Choi, Y., Borghesani, P. R., Chan, J. A., and Segal, R. A. (2005). Migration from a Mitogenic Niche Promotes Cell-Cycle Exit. *J Neurosci 25*, 10437-10445.

Coffer, P. J., Jin, J., and Woodgett, J. R. (1998). Protein kinase B (c-Akt): a multifunctional mediator of phosphatidylinositol 3-kinase activation. *Biochem J 335*, 1-13.

Conover, J. C., Doetsch, F., Garcia-Verdugo, J. M., Gale, N. W., Yancopoulos, G. D., and Alvarez-Buylla, A. (2000). Disruption of Eph/ephrin signaling affects migration and proliferation in the adult subventricular zone. *Nat Neurosci 3*, 1091-1097.

Corotto, F. S., Henegar, J. R., and Maruniak, J. A. (1994). Odor deprivation leads to reduced neurogenesis and reduced neuronal survival in the olfactory bulb of the adult mouse. *Neuroscience 61*, 739-744.

Coskun, V., and Luskin, M. B. (2002). Intrinsic and extrinsic regulation of the proliferation and differentiation of cells in the rodent rostral migratory stream. *J Neurosci Res 69*, 795-802.

Cowan, C. W., Shao, Y. R., Sahin, M., Shamah, S. M., Lin, M. Z., Greer, P. L., Gao, S., Griffith, E. C., Brugge, J. S., and Greenberg, M. E. (2005). Vav family GEFs link activated Ephs to endocytosis and axon guidance. *Neuron 46*, 205-217.

Cremer, H., Lange, R., Christoph, A., Plomann, M., Vopper, G., Roes, J., Brown, R., Baldwin, S., Kraemer, P., Scheff, S., and et al. (1994). Inactivation of the N-CAM gene in mice results in size reduction of the olfactory bulb and deficits in spatial learning. *Nature 367*, 455-459.

Cross, D. A., Alessi, D. R., Cohen, P., Andjelkovich, M., and Hemmings, B. A. (1995). Inhibition of glycogen synthase kinase-3 by insulin mediated by protein kinase B. *Nature 378*, 785-789.

Daniel, J. M., and Reynolds, A. B. (1995). The tyrosine kinase substrate p120cas binds directly to E-cadherin but not to the adenomatous polyposis coli protein or alpha-catenin. *Mol Cell Biol 15*, 4819-4824.

Davy, A., Gale, N. W., Murray, E. W., Klinghoffer, R. A., Soriano, P., Feuerstein, C., and Robbins, S. M. (1999). Compartmentalized signaling by GPI-anchored ephrin-A5 requires the Fyn tyrosine kinase to regulate cellular adhesion. *Genes Dev 13*, 3125-3135.

Davy, A., and Robbins, S. M. (2000). Ephrin-A5 modulates cell adhesion and morphology in an integrin-dependent manner. *EMBO J 19*, 5396-5405.

de Wit, J., and Verhaagen, J. (2003). Role of semaphorins in the adult nervous system. *Prog Neurobiol 71*, 249-267.

Dedhar, S. (2000). Cell-substrate interactions and signaling through ILK. *Curr Opin Cell Biol 12*, 250-256.

del Pozo, M. A., Alderson, N. B., Kiosses, W. B., Chiang, H. H., Anderson, R. G., and Schwartz, M. A. (2004). Integrins regulate Rac targeting by internalization of membrane domains. *Science 303*, 839-842.

Del Pozo, M. A., Kiosses, W. B., Alderson, N. B., Meller, N., Hahn, K. M., and Schwartz, M. A. (2002). Integrins regulate GTP-Rac localized effector interactions through dissociation of Rho-GDI. *Nat Cell Biol 4*, 232-239.

Delcommenne, M., Tan, C., Gray, V., Rue, L., Woodgett, J., and Dedhar, S. (1998). Phosphoinositide-3-OH kinase-dependent regulation of glycogen synthase kinase 3 and protein kinase B/AKT by the integrin-linked kinase. *Proc Natl Acad Sci U S A 95*, 11211-11216.

Dickson, T. C., Mintz, C. D., Benson, D. L., and Salton, S. R. (2002). Functional binding interaction identified between the axonal CAM L1 and members of the ERM family. *J Cell Biol 157*, 1105-1112.

Doetsch, F., Caille, I., Lim, D. A., Garcia-Verdugo, J. M., and Alvarez-Buylla, A. (1999). Subventricular zone astrocytes are neural stem cells in the adult mammalian brain. *Cell 97*, 703-716.

Downward, J. (1998). Mechanisms and consequences of activation of protein kinase B/Akt. *Curr Opin Cell Biol 10*, 262-267.

Dulabon, L., Olson, E. C., Taglienti, M. G., Eisenhuth, S., McGrath, B., Walsh, C. A., Kreidberg, J. A., and Anton, E. S. (2000). Reelin binds alpha3beta1 integrin and inhibits neuronal migration. *Neuron 27*, 33-44.

Edwards, D. C., Sanders, L. C., Bokoch, G. M., and Gill, G. N. (1999). Activation of LIM-kinase by Pak1 couples Rac/Cdc42 GTPase signalling to actin cytoskeletal dynamics. *Nat Cell Biol 1*, 253-259.

Emsley, J. G., and Hagg, T. (2003). alpha6beta1 integrin directs migration of neuronal precursors in adult mouse forebrain. *Exp Neurol 183*, 273-285.

Engelkamp, D. (2002). Cloning of three mouse Unc5 genes and their expression patterns at mid-gestation. *Mech Dev 118*, 191-197.

Eriksson, P. S., Perfilieva, E., Bjork-Eriksson, T., Alborn, A. M., Nordborg, C., Peterson, D. A., and Gage, F. H. (1998). Neurogenesis in the adult human hippocampus. *Nat Med 4*, 1313-1317.

Falls, D. L. (2003). Neuregulins: functions, forms, and signaling strategies. *Exp Cell Res 284*, 14-30.

Flames, N., Long, J. E., Garratt, A. N., Fischer, T. M., Gassmann, M., Birchmeier, C., Lai, C., Rubenstein, J. L., and Marin, O. (2004). Short- and long-range attraction of cortical GABAergic interneurons by neuregulin-1. *Neuron 44*, 251-261.

Flanagan, J. G., and Vanderhaeghen, P. (1998). The ephrins and Eph receptors in neural development. *Annu Rev Neurosci* 1998 21, 309-345.

Forster, E., Tielsch, A., Saum, B., Weiss, K. H., Johanssen, C., Graus-Porta, D., Muller, U., and Frotscher, M. (2002). Reelin, Disabled 1, and beta 1 integrins are required for the formation of the radial glial scaffold in the hippocampus. *Proc Natl Acad Sci U S A 99*, 13178-13183.

Fujimoto, I., Bruses, J. L., and Rutishauser, U. (2001). Regulation of cell adhesion by polysialic acid. Effects on cadherin, immunoglobulin cell adhesion molecule, and integrin function and independence from neural cell adhesion molecule binding or signaling activity. *J Biol Chem 276*, 31745-31751.

Fujisawa, H., and Kitsukawa, T. (1998). Receptors for collapsin/semaphorins. *Curr Opin Neurobiol 8*, 587-592.

Gautreau, A., Louvard, D., and Arpin, M. (2002). ERM proteins and NF2 tumor suppressor: the Yin and Yang of cortical actin organization and cell growth signaling. *Curr Opin Cell Biol 14*, 104-109.

Gavard, J., Marthiens, V., Monnet, C., Lambert, M., and Mege, R. M. (2004). N-cadherin activation substitutes for the cell contact control in cell cycle arrest and myogenic differentiation: involvement of p120 and beta-catenin. *J Biol Chem 279*, 36795-36802.

Giachino, C., De Marchis, S., Giampietro, C., Parlato, R., Perroteau, I., Schutz, G., Fasolo, A., and Peretto, P. (2005). cAMP Response Element-Binding Protein Regulates

Differentiation and Survival of Newborn Neurons in the Olfactory Bulb. *J Neurosci 25*, 10105-10118.

Giger, R. J., Pasterkamp, R. J., Heijnen, S., Holtmaat, A. J., and Verhaagen, J. (1998). Anatomical distribution of the chemorepellent semaphorin III/collapsin-1 in the adult rat and human brain: predominant expression in structures of the olfactory-hippocampal pathway and the motor system. *J Neurosci Res 52*, 27-42.

Gould, K. L., Bretscher, A., Esch, F. S., and Hunter, T. (1989). cDNA cloning and sequencing of the protein-tyrosine kinase substrate, ezrin, reveals homology to band 4.1. *EMBO J 8*, 4133-4142.

Grosheva, I., Shtutman, M., Elbaum, M., and Bershadsky, A. D. (2001). p120 catenin affects cell motility via modulation of activity of Rho-family GTPases: a link between cell-cell contact formation and regulation of cell locomotion. *J Cell Sci 114*, 695-707.

Guan, J. L. (2004). Cell biology. Integrins, rafts, Rac, and Rho. *Science 303*, 773-774.

Guan, K. L., and Rao, Y. (2003). Signalling mechanisms mediating neuronal responses to guidance cues. *Nat Rev Neurosci 4*, 941-956.

Hack, I., Bancila, M., Loulier, K., Carroll, P., and Cremer, H. (2002). Reelin is a detachment signal in tangential chain-migration during postnatal neurogenesis. *Nature Neuroscience 5*, 939-945.

Hack, M. A., Saghatelyan, A., de Chevigny, A., Pfeifer, A., Ashery-Padan, R., Lledo, P. M., and Gotz, M. (2005). Neuronal fate determinants of adult olfactory bulb neurogenesis. *Nat Neurosci. 8*, 865–872

Haegele, L., Ingold, B., Naumann, H., Tabatabai, G., Ledermann, B., and Brandner, S. (2003). Wnt signalling inhibits neural differentiation of embryonic stem cells by controlling bone morphogenetic protein expression. *Mol Cell Neurosci 24*, 696-708.

Hall, A. (1998). Rho GTPases and the actin cytoskeleton. *Science 279*, 509-514.

Hartwig, J. H., Bokoch, G. M., Carpenter, C. L., Janmey, P. A., Taylor, L. A., Toker, A., and Stossel, T. P. (1995). Thrombin receptor ligation and activated Rac uncap actin filament barbed ends through phosphoinositide synthesis in permeabilized human platelets. *Cell 82*, 643-653.

Haspel, J., and Grumet, M. (2003). The L1CAM extracellular region: a multi-domain protein with modular and cooperative binding modes. *Front Biosci 8*, s1210-1225.

He, Z., Wang, K. C., Koprivica, V., Ming, G., and Song, H. J. (2002). Knowing how to navigate: mechanisms of semaphorin signaling in the nervous system. *Sci STKE 2002*, RE1.

Hemmings, B. A. (1997). Akt signaling: linking membrane events to life and death decisions. *Science 275*, 628-630.

Hiesberger, T., Trommsdorff, M., Howell, B. W., Goffinet, A., Mumby, M. C., Cooper, J. A., and Herz, J. (1999). Direct binding of Reelin to VLDL receptor and ApoE receptor 2 induces tyrosine phosphorylation of disabled-1 and modulates tau phosphorylation. *Neuron 24*, 481-489.

Hirao, M., Sato, N., Kondo, T., Yonemura, S., Monden, M., Sasaki, T., Takai, Y., and Tsukita, S. (1996). Regulation mechanism of ERM (ezrin/radixin/moesin) protein/plasma membrane association: possible involvement of phosphatidylinositol turnover and Rho-dependent signaling pathway. *J Cell Biol 135*, 37-51.

Hiscox, S., and Jiang, W. G. (1999). Ezrin regulates cell-cell and cell-matrix adhesion, a possible role with E-cadherin/beta-catenin. *J Cell Sci 112 Pt 18*, 3081-3090.

Hu, H. (1999). Chemorepulsion of neuronal migration by Slit2 in the developing mammalian forebrain. *Neuron 23*, 703-711.

Hu, H., Tomasiewicz, H., Magnuson, T., and Rutishauser, U. (1996). The role of polysialic acid in migration of olfactory bulb interneuron precursors in the subventricular zone. *Neuron 16,* 735-743.

Huai, J., and Drescher, U. (2001). An ephrin-A-dependent signaling pathway controls integrin function and is linked to the tyrosine phosphorylation of a 120-kDa protein. *J Biol Chem 276*, 6689-6694.

Hynes, R. O. (2002). Integrins: bidirectional, allosteric signaling machines. Cell 110, 673-687.

Irie, F., and Yamaguchi, Y. (2002). EphB receptors regulate dendritic spine development via intersectin, Cdc42 and N-WASP. *Nat Neurosci 5*, 1117-1118.

Itoh, A., Miyabayashi, T., Ohno, M., and Sakano, S. (1998). Cloning and expressions of three mammalian homologues of Drosophila slit suggest possible roles for Slit in the formation and maintenance of the nervous system. *Brain Res Mol Brain Res 62*, 175-186.

Jacques, T. S., Relvas, J. B., Nishimura, S., Pytela, R., Edwards, G. M., Streuli, C. H., and ffrench-Constant, C. (1998). Neural precursor cell chain migration and division are regulated through different beta1 integrins. *Development 125*, 3167-3177.

Jenkins, S. M., Kizhatil, K., Kramarcy, N. R., Sen, A., Sealock, R., and Bennett, V. (2001). FIGQY phosphorylation defines discrete populations of L1 cell adhesion molecules at sites of cell-cell contact and in migrating neurons. *J Cell Sci 114*, 3823-3835.

Jin, Z., and Strittmatter, S. M. (1997). Rac1 mediates collapsin-1-induced growth cone collapse. *J Neurosci 17*, 6256-6263.

Johnson, M. W., Miyata, H., and Vinters, H. V. (2002). Ezrin and moesin expression within the developing human cerebrum and tuberous sclerosis-associated cortical tubers. *Acta Neuropathol (Berl) 104*, 188-196.

Jones, F. S., and Jones, P. L. (2000). The tenascin family of ECM glycoproteins: structure, function, and regulation during embryonic development and tissue remodeling. *Dev Dyn 218*, 235-259.

Jurney, W. M., Gallo, G., Letourneau, P. C., and McLoon, S. C. (2002). Rac1-mediated endocytosis during ephrin-A2- and semaphorin 3A-induced growth cone collapse. *J Neurosci 22*, 6019-6028.

Kaibuchi, K., Kuroda, S., and Amano, M. (1999). Regulation of the cytoskeleton and cell adhesion by the Rho family GTPases in mammalian cells. *Annu Rev Biochem 68*, 459-486.

Kaverina, I., Krylyshkina, O., and Small, J. V. (2002). Regulation of substrate adhesion dynamics during cell motility. *Int J Biochem Cell Biol 34*, 746-761.

Keino-Masu, K., Masu, M., Hinck, L., Leonardo, E. D., Chan, S. S., Culotti, J. G., and Tessier-Lavigne, M. (1996). Deleted in Colorectal Cancer (DCC) encodes a netrin receptor. *Cell 87*, 175-185.

Kenwrick, S., Watkins, A., and De Angelis, E. (2000). Neural cell recognition molecule L1: relating biological complexity to human disease mutations. *Hum Mol Genet 9*, 879-886.

Klemke, R. L., Cai, S., Giannini, A. L., Gallagher, P. J., de Lanerolle, P., and Cheresh, D. A. (1997). Regulation of cell motility by mitogen-activated protein kinase. *J Cell Biol 137*, 481-492.

Kohwi, M., Osumi, N., Rubenstein, J. L. R., and Alvarez-Buylla, A. (2005). Pax6 Is Required for Making Specific Subpopulations of Granule and Periglomerular Neurons in the Olfactory Bulb. *J Neurosci 25*, 6997-7003.

Kolodkin, A. L., Matthes, D. J., and Goodman, C. S. (1993). The semaphorin genes encode a family of transmembrane and secreted growth cone guidance molecules. *Cell 75*, 1389-1399.

Kornack, D. R., and Rakic, P. (1999). Continuation of neurogenesis in the hippocampus of the adult macaque monkey. *Proc Natl Acad Sci U S A 96*, 5768-5773.

Kuhn, H. G., Dickinson-Anson, H., and Gage, F. H. (1996). Neurogenesis in the dentate gyrus of the adult rat: age-related decrease of neuronal progenitor proliferation. *J Neurosci 16*, 2027-2033.

Kuhn, T. B., Brown, M. D., Wilcox, C. L., Raper, J. A., and Bamburg, J. R. (1999). Myelin and collapsin-1 induce motor neuron growth cone collapse through different pathways: inhibition of collapse by opposing mutants of rac1. *J Neurosci 19*, 1965-1975.

Lambert de Rouvroit, C., and Goffinet, A. M. (1998). A new view of early cortical development. *Biochem Pharmacol 56*, 1403-1409.

Larson, J., Hoffman, J. S., Guidotti, A., and Costa, E. (2003). Olfactory discrimination learning deficit in heterozygous reeler mice. *Brain Res 971*, 40-46.

Lauffenburger, D. A., and Horwitz, A. F. (1996). Cell migration: a physically integrated molecular process. *Cell 84*, 359-369.

LeCouter, J., Lin, R., Tejada, M., Frantz, G., Peale, F., Hillan, K. J., and Ferrara, N. (2003). The endocrine-gland-derived VEGF homologue Bv8 promotes angiogenesis in the testis: Localization of Bv8 receptors to endothelial cells. *Proc Natl Acad Sci U S A 100*, 2685-2690.

Lemasson, M., Saghatelyan, A., Olivo-Marin, J.-C., and Lledo, P.-M. (2005). Neonatal and Adult Neurogenesis Provide Two Distinct Populations of Newborn Neurons to the Mouse Olfactory Bulb. *J Neurosci 25*, 6816-6825.

Leonardo, E. D., Hinck, L., Masu, M., Keino-Masu, K., Ackerman, S. L., and Tessier-Lavigne, M. (1997). Vertebrate homologues of C. elegans UNC-5 are candidate netrin receptors. *Nature 386*, 833-838.

Li, M., Bullock, C. M., Knauer, D. J., Ehlert, F. J., and Zhou, Q. Y. (2001). Identification of two prokineticin cDNAs: recombinant proteins potently contract gastrointestinal smooth muscle. *Mol Pharmacol 59*, 692-698.

Li, S., Guan, J. L., and Chien, S. (2005). Biochemistry and biomechanics of cell motility. *Annu Rev Biomed Eng 7*, 105-150.

Li, W., Lee, J., Vikis, H. G., Lee, S. H., Liu, G., Aurandt, J., Shen, T. L., Fearon, E. R., Guan, J. L., Han, M., et al. (2004). Activation of FAK and Src are receptor-proximal events required for netrin signaling. *Nat Neurosci 7*, 1213-1221.

Li, X., Saint-Cyr-Proulx, E., Aktories, K., and Lamarche-Vane, N. (2002). Rac1 and Cdc42 but not RhoA or Rho kinase activities are required for neurite outgrowth induced by the Netrin-1 receptor DCC (deleted in colorectal cancer) in N1E-115 neuroblastoma cells. *J Biol Chem 277*, 15207-15214.

Liang, J., Zubovitz, J., Petrocelli, T., Kotchetkov, R., Connor, M. K., Han, K., Lee, J. H., Ciarallo, S., Catzavelos, C., Beniston, R., et al. (2002). PKB/Akt phosphorylates p27, impairs nuclear import of p27 and opposes p27-mediated G1 arrest. *Nat Med 8*, 1153-1160.

Lim, D. A., and Alvarez-Buylla, A. (1999). Interaction between astrocytes and adult subventricular zone precursors stimulates neurogenesis. *Proc Natl Acad Sci U S A 96*, 7526-7531.

Lin, D. C., Bullock, C. M., Ehlert, F. J., Chen, J. L., Tian, H., and Zhou, Q. Y. (2002). Identification and molecular characterization of two closely related G protein-coupled receptors activated by prokineticins/endocrine gland vascular endothelial growth factor. *J Biol Chem 277*, 19276-19280.

Liu, G., Beggs, H., Jurgensen, C., Park, H. T., Tang, H., Gorski, J., Jones, K. R., Reichardt, L. F., Wu, J., and Rao, Y. (2004). Netrin requires focal adhesion kinase and Src family kinases for axon outgrowth and attraction. *Nat Neurosci 7,* 1222-1232.

Liu, G., and Rao, Y. (2003). Neuronal migration from the forebrain to the olfactory bulb requires a new attractant persistent in the olfactory bulb. *J Neurosci 23*, 6651-6659.

Liu, X., Wang, Q., Haydar, T. F., and Bordey, A. (2005). Nonsynaptic GABA signaling in postnatal subventricular zone controls proliferation of GFAP-expressing progenitors. *Nat Neurosci 8*, 1179-1187.

Lledo, P. M., and Saghatelyan, A. (2005). Integrating new neurons into the adult olfactory bulb: joining the network, life-death decisions, and the effects of sensory experience. *Trends Neurosci 28*, 248-254.

Lodish , H., Darnell, J., Matsudaira, P., Berk, A., and Zipursky, S. (2003). The Dynamics of Actin Assembly. In *Molecular cell biology* (New York, W.H. Freeman and Company), pp. 784-791.

Logan, C. Y., and Nusse, R. (2004). The Wnt signaling pathway in development and disease. *Annu Rev Cell Dev Biol 20*, 781-810.

Lois, C., and Alvarez-Buylla, A. (1994). Long-distance neuronal migration in the adult mammalian brain. *Science 264*, 1145-1148.

Lois, C., Garcia-Verdugo, J. M., and Alvarez-Buylla, A. (1996). Chain migration of neuronal precursors. *Science 271*, 978-981.

Long, H., Sabatier, C., Ma, L., Plump, A., Yuan, W., Ornitz, D. M., Tamada, A., Murakami, F., Goodman, C. S., and Tessier-Lavigne, M. (2004). Conserved roles for Slit and Robo proteins in midline commissural axon guidance. *Neuron 42,* 213-223.

Louvet-Vallee, S. (2000). ERM proteins: from cellular architecture to cell signaling. *Biol Cell 92*, 305-316.

Luskin, M. B. (1993). Restricted proliferation and migration of postnatally generated neurons derived from the forebrain subventricular zone. *Neuron 11*, 173-189.

Maeda, M., Matsui, T., Imamura, M., and Tsukita, S. (1999). Expression level, subcellular distribution and rho-GDI binding affinity of merlin in comparison with Ezrin/Radixin/Moesin proteins. *Oncogene 18*, 4788-4797.

Maekawa, M., Ishizaki, T., Boku, S., Watanabe, N., Fujita, A., Iwamatsu, A., Obinata, T., Ohashi, K., Mizuno, K., and Narumiya, S. (1999). Signaling from Rho to the actin cytoskeleton through protein kinases ROCK and LIM-kinase. *Science 285*, 895-898.

Marillat, V., Cases, O., Nguyen-Ba-Charvet, K. T., Tessier-Lavigne, M., Sotelo, C., and Chedotal, A. (2002). Spatiotemporal expression patterns of slit and robo genes in the rat brain. *J Comp Neurol 442*, 130-155.

Marin, O., and Rubenstein, J. L. (2001). A long, remarkable journey: tangential migration in the telencephalon. *Nat Rev Neurosci 2*, 780-790.

Marin, O., and Rubenstein, J. L. (2003). Cell migration in the forebrain. *Annu Rev Neurosci 26*, 441-483.

Marin, O., Yaron, A., Bagri, A., Tessier-Lavigne, M., and Rubenstein, J. L. (2001). Sorting of striatal and cortical interneurons regulated by semaphorin-neuropilin interactions. *Science 293*, 872-875.

Mason, H. A., Ito, S., and Corfas, G. (2001). Extracellular signals that regulate the tangential migration of olfactory bulb neuronal precursors: inducers, inhibitors, and repellents. *J Neurosci 21*, 7654-7663.

Matarredona, E. R., Murillo-Carretero, M., Moreno-Lopez, B., and Estrada, C. (2005). Role of nitric oxide in subventricular zone neurogenesis. *Brain Res Brain Res Rev 49*, 355-366.

Menezes, J. R., Smith, C. M., Nelson, K. C., and Luskin, M. B. (1995). The division of neuronal progenitor cells during migration in the neonatal mammalian forebrain. *Mol Cell Neurosci 6*, 496-508.

Miki, H., Yamaguchi, H., Suetsugu, S., and Takenawa, T. (2000). IRSp53 is an essential intermediate between Rac and WAVE in the regulation of membrane ruffling. *Nature 408*, 732-735.

Mintz, C. D., Dickson, T. C., Gripp, M. L., Salton, S. R., and Benson, D. L. (2003). ERMs colocalize transiently with L1 during neocortical axon outgrowth. *J Comp Neurol 464*, 438-448.

Miranti, C. K., and Brugge, J. S. (2002). Sensing the environment: a historical perspective on integrin signal transduction. *Nat Cell Biol 4*, E83-90.

Mitchison, T. J., and Cramer, L. P. (1996). Actin-based cell motility and cell locomotion. *Cell 84*, 371-379.

Murai, K. K., and Pasquale, E. B. (2005). New Exchanges in Eph-Dependent Growth Cone Dynamics. *Neuron 46*, 161-163.

Murase, S., and Horwitz, A. F. (2002). Deleted in colorectal carcinoma and differentially expressed integrins mediate the directional migration of neural precursors in the rostral migratory stream. *J Neurosci 22*, 3568-3579.

Negishi, M., Oinuma, I., and Katoh, H. (2005). Plexins: axon guidance and signal transduction. *Cell Mol Life Sci 62*, 1363-1371.

Nelson, W. J., and Nusse, R. (2004). Convergence of Wnt, {beta}-Catenin, and Cadherin Pathways. *Science 303*, 1483-1487.

Ng, K. L., Li, J. D., Cheng, M. Y., Leslie, F. M., Lee, A. G., and Zhou, Q. Y. (2005). Dependence of olfactory bulb neurogenesis on prokineticin 2 signaling. *Science 308*, 1923-1927.

Nguyen, D. H., Catling, A. D., Webb, D. J., Sankovic, M., Walker, L. A., Somlyo, A. V., Weber, M. J., and Gonias, S. L. (1999). Myosin light chain kinase functions downstream of Ras/ERK to promote migration of urokinase-type plasminogen activator-stimulated cells in an integrin-selective manner. *J Cell Biol 146*, 149-164.

Nguyen-Ba-Charvet, K. T., Picard-Riera, N., Tessier-Lavigne, M., Baron-Van Evercooren, A., Sotelo, C., and Chedotal, A. (2004). Multiple roles for slits in the control of cell migration in the rostral migratory stream. *J Neurosci 24*, 1497-1506.

Ni, C. Y., Murphy, M. P., Golde, T. E., and Carpenter, G. (2001). gamma -Secretase cleavage and nuclear localization of ErbB-4 receptor tyrosine kinase. *Science 294*, 2179-2181.

Noren, N. K., Liu, B. P., Burridge, K., and Kreft, B. (2000). p120 catenin regulates the actin cytoskeleton via Rho family GTPases. *J Cell Biol 150*, 567-580.

Noren, N. K., and Pasquale, E. B. (2004). Eph receptor-ephrin bidirectional signals that target Ras and Rho proteins. *Cell Signal 1*6, 655-666.

O'Leary, D. D., and Wilkinson, D. G. (1999). Eph receptors and ephrins in neural development. *Curr Opin Neurobiol 9*, 65-73.

O'Rourke, N. A. (1996). Neuronal chain gangs: homotypic contacts support migration into the olfactory bulb. *Neuron 16*, 1061-1064.

Ono, K., Tomasiewicz, H., Magnuson, T., and Rutishauser, U. (1994). N-CAM mutation inhibits tangential neuronal migration and is phenocopied by enzymatic removal of polysialic acid. *Neuron 13,* 595-609.

Palazzo, A. F., Cook, T. A., Alberts, A. S., and Gundersen, G. G. (2001). mDia mediates Rho-regulated formation and orientation of stable microtubules. *Nat Cell Biol 3*, 723-729.

Palma, V., Lim, D. A., Dahmane, N., Sanchez, P., Brionne, T. C., Herzberg, C. D., Gitton, Y., Carleton, A., Alvarez-Buylla, A., and Ruiz i Altaba, A. (2005). Sonic hedgehog controls stem cell behavior in the postnatal and adult brain. *Development 132*, 335-344.

Palmer, A., and Klein, R. (2003). Multiple roles of ephrins in morphogenesis, neuronal networking, and brain function. *Genes Dev 17*, 1429-1450.

Palmer, A., Zimmer, M., Erdmann, K. S., Eulenburg, V., Porthin, A., Heumann, R., Deutsch, U., and Klein, R. (2002). EphrinB phosphorylation and reverse signaling: regulation by Src kinases and PTP-BL phosphatase. *Mol Cell 9*, 725-737.

Palmer, T. D., Willhoite, A. R., and Gage, F. H. (2000). Vascular niche for adult hippocampal neurogenesis. *J Comp Neurol 425*, 479-494.

Pappas, G. D., Kriho, V., Liu, W. S., Tremolizzo, L., Lugli, G., and Larson, J. (2003). Immunocytochemical localization of reelin in the olfactory bulb of the heterozygous reeler mouse: an animal model for schizophrenia. *Neurol Res 25*, 819-830.

Park, H. T., Wu, J., and Rao, Y. (2002). Molecular control of neuronal migration. *Bioessays 24*, 821-827.

Parrini, M. C., Lei, M., Harrison, S. C., and Mayer, B. J. (2002). Pak1 kinase homodimers are autoinhibited in trans and dissociated upon activation by Cdc42 and Rac1. *Mol Cell 9*, 73-83.

Pasterkamp, R. J., and Kolodkin, A. L. (2003). Semaphorin junction: making tracks toward neural connectivity. *Curr Opin Neurobiol 13*, 79-89.

Penzes, P., Beeser, A., Chernoff, J., Schiller, M. R., Eipper, B. A., Mains, R. E., and Huganir, R. L. (2003). Rapid induction of dendritic spine morphogenesis by trans-synaptic ephrinB-EphB receptor activation of the Rho-GEF kalirin. *Neuron 37*, 263-274.

Peretto, P., Merighi, A., Fasolo, A., and Bonfanti, L. (1997). Glial tubes in the rostral migratory stream of the adult rat. *Brain Res Bull 42*, 9-21.

Perrot, V., Vazquez-Prado, J., and Gutkind, J. S. (2002). Plexin B regulates Rho through the guanine nucleotide exchange factors leukemia-associated Rho GEF (LARG) and PDZ-RhoGEF. *J Biol Chem 277*, 43115-43120.

Pesheva, P., and Probstmeier, R. (2000). The yin and yang of tenascin-R in CNS development and pathology. *Prog Neurobiol 61*, 465-493.

Petridis, A. K., El-Maarouf, A., and Rutishauser, U. (2004). Polysialic acid regulates cell contact-dependent neuronal differentiation of progenitor cells from the subventricular zone. *Dev Dyn 230*, 675-684.

Pollard, T. D., Blanchoin, L., and Mullins, R. D. (2000). Molecular mechanisms controlling actin filament dynamics in nonmuscle cells. Annu Rev Biophys Biomol Struct 29, 545-576.

Pollard, T. D., and Borisy, G. G. (2003). Cellular motility driven by assembly and disassembly of actin filaments. *Cell 112*, 453-465.

Ponti, A., Machacek, M., Gupton, S. L., Waterman-Storer, C. M., and Danuser, G. (2004). Two Distinct Actin Networks Drive the Protrusion of Migrating Cells. *Science 305*, 1782-1786.

Porfiri, E., Rubinfeld, B., Albert, I., Hovanes, K., Waterman, M., and Polakis, P. (1997). Induction of a beta-catenin-LEF-1 complex by wnt-1 and transforming mutants of beta-catenin. *Oncogene 15*, 2833-2839.

Przyborski, S. A., Knowles, B. B., and Ackerman, S. L. (1998). Embryonic phenotype of Unc5h3 mutant mice suggests chemorepulsion during the formation of the rostral cerebellar boundary. *Development 125*, 41-50.

Rafelski, S. M., and Theriot, J. A. (2004). Crawling toward a unified model of cell mobility: spatial and temporal regulation of actin dynamics. *Annu Rev Biochem 73*, 209-239.

Raftopoulou, M., and Hall, A. (2004). Cell migration: Rho GTPases lead the way. *Dev Biol 265*, 23-32.

Rakic, P., and Caviness, V. S., Jr. (1995). Cortical development: view from neurological mutants two decades later. *Neuron 14*, 1101-1104.

Redies, C. (2000). Cadherins in the central nervous system. *Prog Neurobiol 61*, 611-648.

Ren, X. R., Ming, G. L., Xie, Y., Hong, Y., Sun, D. M., Zhao, Z. Q., Feng, Z., Wang, Q., Shim, S., Chen, Z. F., et al. (2004). Focal adhesion kinase in netrin-1 signaling. *Nat Neurosci 7*, 1204-1212.

Reynolds, A. B., Daniel, J., McCrea, P. D., Wheelock, M. J., Wu, J., and Zhang, Z. (1994). Identification of a new catenin: the tyrosine kinase substrate p120cas associates with E-cadherin complexes. *Mol Cell Biol 14*, 8333-8342.

Rhee, J., Mahfooz, N. S., Arregui, C., Lilien, J., Balsamo, J., and VanBerkum, M. F. (2002). Activation of the repulsive receptor Roundabout inhibits N-cadherin-mediated cell adhesion. *Nat Cell Biol 4*, 798-805.

Rice, D. S., and Curran, T. (2001). Role of the reelin signaling pathway in central nervous system development. *Annu Rev Neurosci 24*, 1005-1039.

Ridley, A. J. (2001). Rho GTPases and cell migration. *J Cell Sci 114*, 2713-2722.

Ridley, A. J., Schwartz, M. A., Burridge, K., Firtel, R. A., Ginsberg, M. H., Borisy, G., Parsons, J. T., and Horwitz, A. R. (2003). Cell migration: integrating signals from front to back. *Science 302*, 1704-1709.

Riese, D. J., 2nd, and Stern, D. F. (1998). Specificity within the EGF family/ErbB receptor family signaling network. *Bioessays 20*, 41-48.

Rio, C., Buxbaum, J. D., Peschon, J. J., and Corfas, G. (2000). Tumor necrosis factor-alpha-converting enzyme is required for cleavage of erbB4/HER4. *J Biol Chem 275*, 10379-10387.

Rousselot, P., Lois, C., and Alvarez-Buylla, A. (1995). Embryonic (PSA) N-CAM reveals chains of migrating neuroblasts between the lateral ventricle and the olfactory bulb of adult mice. *J Comp Neurol 351*, 51-61.

Rugarli, E. I., and Ballabio, A. (1995). Reelin: a novel extracellular matrix protein involved in brain lamination. *Bioessays 17*, 832-834.

Saghatelyan, A., de Chevigny, A., Schachner, M., and Lledo, P. M. (2004). Tenascin-R mediates activity-dependent recruitment of neuroblasts in the adult mouse forebrain. *Nat Neurosci 7*, 347-356.

Sahin, M., Greer, P. L., Lin, M. Z., Poucher, H., Eberhart, J., Schmidt, S., Wright, T. M., Shamah, S. M., O'Connell, S., Cowan, C. W., et al. (2005). Eph-dependent tyrosine phosphorylation of ephexin1 modulates growth cone collapse. *Neuron 46*, 191-204.

Sanders, L. C., Matsumura, F., Bokoch, G. M., and de Lanerolle, P. (1999). Inhibition of myosin light chain kinase by p21-activated kinase. *Science 283*, 2083-2085.

Schaller, M. D. (2001). Biochemical signals and biological responses elicited by the focal adhesion kinase. *Biochim Biophys Acta 1540*, 1-21.

Schmid, R. S., and Anton, E. S. (2003). Role of integrins in the development of the cerebral cortex. *Cereb Cortex 13*, 219-224.

Schmid, R. S., Jo, R., Shelton, S., Kreidberg, J. A., and Anton, E. S. (2005). Reelin, Integrin and Dab1 Interactions during Embryonic Cerebral Cortical Development. *Cereb Cortex 15*, 1632-1636.

Schmidt, A., and Hall, A. (2002). Guanine nucleotide exchange factors for Rho GTPases: turning on the switch. *Genes Dev 16*, 1587-1609.

Serafini, T., Colamarino, S. A., Leonardo, E. D., Wang, H., Beddington, R., Skarnes, W. C., and Tessier-Lavigne, M. (1996). Netrin-1 is required for commissural axon guidance in the developing vertebrate nervous system. *Cell 87*, 1001-1014.

Serini, G., Valdembri, D., Zanivan, S., Morterra, G., Burkhardt, C., Caccavari, F., Zammataro, L., Primo, L., Tamagnone, L., Logan, M., et al. (2003). Class 3 semaphorins control vascular morphogenesis by inhibiting integrin function. *Nature 424*, 391-397.

Shamah, S. M., Lin, M. Z., Goldberg, J. L., Estrach, S., Sahin, M., Hu, L., Bazalakova, M., Neve, R. L., Corfas, G., Debant, A., and Greenberg, M. E. (2001). EphA receptors regulate growth cone dynamics through the novel guanine nucleotide exchange factor ephexin. *Cell 105*, 233-244.

Sheetz, M. P., Felsenfeld, D., Galbraith, C. G., and Choquet, D. (1999). Cell migration as a five-step cycle. *Biochem Soc Symp 65*, 233-243.

Shekarabi, M., and Kennedy, T. E. (2002). The netrin-1 receptor DCC promotes filopodia formation and cell spreading by activating Cdc42 and Rac1. *Mol Cell Neurosci 19*, 1-17.

Sheldon, M., Rice, D. S., D'Arcangelo, G., Yoneshima, H., Nakajima, K., Mikoshiba, K., Howell, B. W., Cooper, J. A., Goldowitz, D., and Curran, T. (1997). Scrambler and yotari disrupt the disabled gene and produce a reeler-like phenotype in mice. *Nature 389*, 730-733.

Shen, Q., Goderie, S. K., Jin, L., Karanth, N., Sun, Y., Abramova, N., Vincent, P., Pumiglia, K., and Temple, S. (2004). Endothelial cells stimulate self-renewal and expand neurogenesis of neural stem cells. *Science 304*, 1338-1340.

Shen, T. L., and Guan, J. L. (2001). Differential regulation of cell migration and cell cycle progression by FAK complexes with Src, PI3K, Grb7 and Grb2 in focal contacts. *FEBS Lett 499, 176-181.*

Shepard, G., and Greer, C. (1998). Olfactory Bulb. In *The synaptic organization of the brain*, G. Shepard, ed. (New York, Oxford University Press), pp. 159-204.

Shibamoto, S., Hayakawa, M., Takeuchi, K., Hori, T., Miyazawa, K., Kitamura, N., Johnson, K. R., Wheelock, M. J., Matsuyoshi, N., Takeichi, M., and et al. (1995). Association of p120, a tyrosine kinase substrate, with E-cadherin/catenin complexes. *J Cell Biol 128*, 949-957.

Small, J. V., and Kaverina, I. (2005). Polarized Cell Motility: Microtubules show the way. In *Cell migration in development and disease*. D. Wedlich, ed. (Weinheim, Wiley-VCH), pp. 15-31.

Soga, T., Matsumoto, S., Oda, T., Saito, T., Hiyama, H., Takasaki, J., Kamohara, M., Ohishi, T., Matsushime, H., and Furuichi, K. (2002). Molecular cloning and characterization of prokineticin receptors. *Biochim Biophys Acta 1579*, 173-179.

Song, H., Stevens, C. F., and Gage, F. H. (2002). Astroglia induce neurogenesis from adult neural stem cells. *Nature 417*, 39-44.

Spradling, A., Drummond-Barbosa, D., and Kai, T. (2001). Stem cells find their niche. *Nature 414*, 98-104.

Staddon, J. M., Smales, C., Schulze, C., Esch, F. S., and Rubin, L. L. (1995). p120, a p120-related protein (p100), and the cadherin/catenin complex. *J Cell Biol 130*, 369-381.

Stefansson, H., Sarginson, J., Kong, A., Yates, P., Steinthorsdottir, V., Gudfinnsson, E., Gunnarsdottir, S., Walker, N., Petursson, H., Crombie, C., et al. (2003). Association of neuregulin 1 with schizophrenia confirmed in a Scottish population. *Am J Hum Genet 72*, 83-87.

Stefansson, H., Sigurdsson, E., Steinthorsdottir, V., Bjornsdottir, S., Sigmundsson, T., Ghosh, S., Brynjolfsson, J., Gunnarsdottir, S., Ivarsson, O., Chou, T. T., et al. (2002). Neuregulin 1 and susceptibility to schizophrenia. *Am J Hum Genet 71*, 877-892.

Stein, E., and Tessier-Lavigne, M. (2001). Hierarchical organization of guidance receptors: silencing of netrin attraction by slit through a Robo/DCC receptor complex. *Science 291*, 1928-1938.

Stull, J. T., Tansey, M. G., Tang, D. C., Word, R. A., and Kamm, K. E. (1993). Phosphorylation of myosin light chain kinase: a cellular mechanism for Ca2+ desensitization. *Mol Cell Biochem* 127-128, 229-237.

Swiercz, J. M., Kuner, R., Behrens, J., and Offermanns, S. (2002). Plexin-B1 directly interacts with PDZ-RhoGEF/LARG to regulate RhoA and growth cone morphology. *Neuron 35*, 51-63.

Symons, M., and Settleman, J. (2000). Rho family GTPases: more than simple switches. *Trends Cell Biol 10*, 415-419.

Takahashi, K., Sasaki, T., Mammoto, A., Takaishi, K., Kameyama, T., Tsukita, S., and Takai, Y. (1997). Direct interaction of the Rho GDP dissociation inhibitor with ezrin/radixin/moesin initiates the activation of the Rho small G protein. *J Biol Chem 272*, 23371-23375.

Takahashi, T., Fournier, A., Nakamura, F., Wang, L. H., Murakami, Y., Kalb, R. G., Fujisawa, H., and Strittmatter, S. M. (1999). Plexin-neuropilin-1 complexes form functional semaphorin-3A receptors. *Cell 99*, 59-69.

Takeuchi, K., Kawashima, A., Nagafuchi, A., and Tsukita, S. (1994). Structural diversity of band 4.1 superfamily members. *J Cell Sci 107*, 1921-1928.

Tamagnone, L., and Comoglio, P. M. (2000). Signalling by semaphorin receptors: cell guidance and beyond. *Trends Cell Biol 10*, 377-383.

Tamagnone, L., and Comoglio, P. M. (2004). To move or not to move? Semaphorin signalling in cell migration. *EMBO Rep 5*, 356-361.

Tamamaki, N., Fujimori, K., Nojyo, Y., Kaneko, T., and Takauji, R. (2003). Evidence that Sema3A and Sema3F regulate the migration of GABAergic neurons in the developing neocortex. *J Comp Neurol 455*, 238-248.

Tanaka, M., Ohashi, R., Nakamura, R., Shinmura, K., Kamo, T., Sakai, R., and Sugimura, H. (2004). Tiam1 mediates neurite outgrowth induced by ephrin-B1 and EphA2. *EMBO J 23*, 1075-1088.

Tate, M. C., Garcia, A. J., Keselowsky, B. G., Schumm, M. A., Archer, D. R., and LaPlaca, M. C. (2004). Specific beta1 integrins mediate adhesion, migration, and differentiation of neural progenitors derived from the embryonic striatum. *Mol Cell Neurosci 27*, 22-31.

Tessier-Lavigne, M., and Goodman, C. S. (1996). The molecular biology of axon guidance. *Science 274*, 1123-1133.

Thelen, K., Kedar, V., Panicker, A. K., Schmid, R. S., Midkiff, B. R., and Maness, P. F. (2002). The neural cell adhesion molecule L1 potentiates integrin-dependent cell migration to extracellular matrix proteins. *J Neurosci 22*, 4918-4931.

Tomasiewicz, H., Ono, K., Yee, D., Thompson, C., Goridis, C., Rutishauser, U., and Magnuson, T. (1993). Genetic deletion of a neural cell adhesion molecule variant (N-CAM-180) produces distinct defects in the central nervous system. *Neuron 11*, 1163-1174.

Tonchev, A. B., Yamashima, T., Sawamoto, K., and Okano, H. (2005). Enhanced proliferation of progenitor cells in the subventricular zone and limited neuronal production in the striatum and neocortex of adult macaque monkeys after global cerebral ischemia. *J Neurosci Res.*

Vanhaesebroeck, B., and Alessi, D. R. (2000). The PI3K-PDK1 connection: more than just a road to PKB. *Biochem J 346*, 561-576.

Viglietto, G., Motti, M. L., Bruni, P., Melillo, R. M., D'Alessio, A., Califano, D., Vinci, F., Chiappetta, G., Tsichlis, P., Bellacosa, A., et al. (2002). Cytoplasmic relocalization and inhibition of the cyclin-dependent kinase inhibitor p27(Kip1) by PKB/Akt-mediated phosphorylation in breast cancer. *Nat Med 8,* 1136-1144.

Wahl, S., Barth, H., Ciossek, T., Aktories, K., and Mueller, B. K. (2000). Ephrin-A5 induces collapse of growth cones by activating Rho and Rho kinase. *J Cell Biol 149*, 263-270.

Wang, D. D., Krueger, D. D., and Bordey, A. (2003). GABA depolarizes neuronal progenitors of the postnatal subventricular zone via GABAA receptor activation. *J Physiol 550*, 785-800.

Wang, T. W., Zhang, H., and Parent, J. M. (2005). Retinoic acid regulates postnatal neurogenesis in the murine subventricular zone-olfactory bulb pathway. *Development.*

Ware, M. L., Fox, J. W., Gonzalez, J. L., Davis, N. M., Lambert de Rouvroit, C., Russo, C. J., Chua, S. C., Jr., Goffinet, A. M., and Walsh, C. A. (1997). Aberrant splicing of a mouse disabled homolog, mdab1, in the scrambler mouse. *Neuron 19*, 239-249.

Watanabe, T., Noritake, J., and Kaibuchi, K. (2005). Regulation of microtubules in cell migration. *Trends Cell Biol 15*, 76-83.

Waterman-Storer, C. M., and Salmon, E. (1999). Positive feedback interactions between microtubule and actin dynamics during cell motility. *Curr Opin Cell Biol 11*, 61-67.

Welch, M. D., and Mullins, R. D. (2002). Cellular control of actin nucleation. *Annu Rev Cell Dev Biol 18*, 247-288.

Wichterle, H., Garcia-Verdugo, J. M., and Alvarez-Buylla, A. (1997). Direct evidence for homotypic, glia-independent neuronal migration. *Neuron 18,* 779-791.

Wiesner, S., Legate, K. R., and Fassler, R. (2005). Integrin-actin interactions. *Cell Mol Life Sci 62*, 1081-1099.

Wilkinson, D. G. (2001). Multiple roles of EPH receptors and ephrins in neural development. *Nat Rev Neurosci 2*, 155-164.

Winner, B., Cooper-Kuhn, C. M., Aigner, R., Winkler, J., and Kuhn, H. G. (2002). Long-term survival and cell death of newly generated neurons in the adult rat olfactory bulb. *Eur J Neurosci 16*, 1681-1689.

Wolf, N. S., and Trentin, J. J. (1968). Hemopoietic colony studies. V. Effect of hemopoietic organ stroma on differentiation of pluripotent stem cells. *J Exp Med 127*, 205-214.

Wong, K., Park, H. T., Wu, J. Y., and Rao, Y. (2002). Slit proteins: molecular guidance cues for cells ranging from neurons to leukocytes. *Curr Opin Genet Dev 12*, 583-591.

Wong, K., Ren, X. R., Huang, Y. Z., Xie, Y., Liu, G., Saito, H., Tang, H., Wen, L., Brady-Kalnay, S. M., Mei, L., et al. (2001). Signal transduction in neuronal migration: roles of GTPase activating proteins and the small GTPase Cdc42 in the Slit-Robo pathway. *Cell 107*, 209-221.

Wu, W., Wong, K., Chen, J., Jiang, Z., Dupuis, S., Wu, J. Y., and Rao, Y. (1999). Directional guidance of neuronal migration in the olfactory system by the protein Slit. *Nature 400*, 331-336.

Wurmser, A. E., Palmer, T. D., and Gage, F. H. (2004). Cellular interactions in the stem cell niche. *Science 304*, 1253-1255.

Yang, S. Z., Kohno, N., Yokoyama, A., Kondo, K., Hamada, H., and Hiwada, K. (2001). Decreased E-cadherin augments beta-catenin nuclear localization: studies in breast cancer cell lines. *Int J Oncol 18*, 541-548.

Yarden, Y., and Sliwkowski, M. X. (2001). Untangling the ErbB signalling network. *Nat Rev Mol Cell Biol 2*, 127-137.

Yau, H. J., Wang, H. F., Lai, C., and Liu, F. C. (2003). Neural development of the neuregulin receptor ErbB4 in the cerebral cortex and the hippocampus: preferential expression by interneurons tangentially migrating from the ganglionic eminences. *Cereb Cortex 13*, 252-264.

Yu, T. W., and Bargmann, C. I. (2001). Dynamic regulation of axon guidance. *Nat Neurosci 4*, 1169-1176.

Zamir, E., and Geiger, B. (2001). Molecular complexity and dynamics of cell-matrix adhesions. *J Cell Sci 114*, 3583-3590.

Zhao, X., Shi, Y., Tang, J., Tang, R., Yu, L., Gu, N., Feng, G., Zhu, S., Liu, H., Xing, Y., et al. (2004). A case control and family based association study of the neuregulin1 gene and schizophrenia. *J Med Genet 41*, 31-34.

Zhou, W., and Carpenter, G. (2000). Heregulin-dependent trafficking and cleavage of ErbB-4. *J Biol Chem 275*, 34737-34743.

Zhu, Y., Li, H., Zhou, L., Wu, J. Y., and Rao, Y. (1999). Cellular and molecular guidance of GABAergic neuronal migration from an extracortical origin to the neocortex. *Neuron 23*, 473-485.

Zrihan-Licht, S., Avraham, S., Jiang, S., Fu, Y., and Avraham, H. K. (2004). Coupling of RAFTK/Pyk2 kinase with c-Abl and their role in the migration of breast cancer cells. *Int J Oncol 24*, 153-159.

NOTES ADDED IN PROOF

Data referred to as Cleary et al., unpublished data, are now in press [1].

A recent paper by Sawamoto and colleagues [2] demonstrates that migration of neuroblasts around the lateral ventricles parallels the flow of cerebrospinal fluid (CSF). The beating of ependymal cilia was required for normal CSF flow, neuronal migration, and the formation of a gradient of the chemorepulsive factor slit2 (discussed above). The authors suggest that creating a gradient of slit protein in the overlying SVZ, the flow of CSF in the lateral ventricles may provide important directional information to cells before they reach the RMS.

Two recent papers by Nelson, Weiss and colleagues [3, 4] provide further insight into the role of the cadherin and catenin families of proteins in regulating actin filament assembly. In the first of these papers [4] they show that contrary to what had been believed previously, α-catenin is unable to bind simultaneously to actin and to a membrane complex containing β-catenin and E-cadherin. In the second paper [3] they show that α-catenin suppresses Arp 2/3 mediated actin polymerization (see figure 3), probably by competing with Arp 2/3 for binding to actin. Interestingly, binding of α-catenin to actin was inhibited by binding of α-catenin to a membrane complex containing β-catenin and E-cadherin.

A review of the role of extracellular matrix proteins in neuronal migration during cerebellar development has been published [5] and offers some interesting parallels and contrasts with the findings described here for the adult SVZ and RMS.

An excellent review of reelin-related signalling pathways as well as a comprehensive description of the reeler phenotype has recently been published [6]. Two recent studies also provide additional insight into the mechanism of reelin signalling in migrating neurons. Simo and colleagues [7] found that Reelin triggers ERK signaling in an SFK/mDab1- and PI3K-dependent manner and that ERK activation is required for Reelin-dependent detachment of neurons migrating from the SVZ, while Oshima and colleagues [8] showed that Cdk5 modulates Reelin signaling through the Ser/Thr phosphorylation of Dab1 in an isoform-specific manner.

An article by Wilson and colleagues [9] shows that in addition to their roles in axon guidance and neuronal migration (reviewed above) netrins also stimulate proliferation, migration and tube formation in endothelial cells.

[1] Cleary, M.A., et al., *Expression of Ezrin in Glial Tubes in the Adult Subventricular Zone and Rostral Migratory Stream.* Neuroscience, 2006. in press.

[2] Sawamoto, K., et al., *New neurons follow the flow of cerebrospinal fluid in the adult brain.* Science, 2006. 311(5761): p. 629-32.

[3] Drees, F., et al., *Alpha-catenin is a molecular switch that binds E-cadherin-beta-catenin and regulates actin-filament assembly.* Cell, 2005. 123(5): p. 903-15.

[4] Yamada, S., et al., *Deconstructing the cadherin-catenin-actin complex.* Cell, 2005. 123(5): p. 889-901.

[5] Porcionatto, M.A., *The extracellular matrix provides directional cues for neuronal migration during cerebellar development.* Braz J Med Biol Res, 2006. 39(3): p. 313-20. Epub 2006 Feb 22.

[6] D'Arcangelo, G., *The reeler mouse: anatomy of a mutant.* Int Rev Neurobiol, 2005. 71: p. 383-417.

[7] Simo, S., et al., *Reelin Induces the Detachment of Postnatal Subventricular Zone Cells and the Expression of the Egr-1 through Erk1/2 Activation.* Cereb Cortex, 2006.

[8] Ohshima, T., et al., *Modulation of Reelin signaling by Cyclin-dependent kinase 5.* Brain Res., 2006. [Epub ahead of print].

[9] Wilson, B.D., et al., *Netrins Promote Developmental and Therapeutic Angiogenesis.* Science, 2006. 313(5787): p. 640-644.

In: Neuronal Migration
Editors: Andrew Girard and Luca Moreau
ISBN: 978-1-62100-059-4

Chapter V

Neurogenesis and Neural Progenitor Migration after Stroke

Monica J. Chau, Ian L. McCullough, Shan Ping Yu, and Ling Wei*
Department of Anesthesiology, Emory University School of Medicine
Atlanta, Georgia, US

Abstract

Ischemic stroke is a leading cause of death and long-term disability worldwide. However, repairing damaged brain tissue is not feasible with current treatments. In recent years, it became clear that specific regions of the adult brain such as the subventricular zone (SVZ) and the subgranular zone (SGZ) are capable of responding to various stimuli including injurious signals from other brain areas by increasing the proliferation of resident neural progenitors. Progenitor proliferation and the migration of the newly formed neural progenitors towards the injured sites are regarded as an important endogenous repair process although current evidence indicates that the proliferation and migration activities are far from sufficient for satisfactory tissue repair. Thus, understanding the endogenous regenerative mechanism including proliferation and migration of neural progenitors is imperative for promoting tissue repair and regenerating an ischemia-injured brain structure.

I. Introduction

Stroke is currently a leading cause of worldwide mortality as well as the leading cause of permanent disability among adults [1]. Approximately 85% of strokes are ischemic in nature and 15% are hemorrhagic [1]. Ischemic stroke is a neurological dysfunction characterized by

* Corresponding author: Ling Wei, MD, Department of Anesthesiology, Emory University School of Medicine, Atlanta, GA 30322, Tel. 404-718-8661, E-mail: lwei7@emory.edu

brain tissue and functional damage resulting from a sudden decrease in cerebral perfusion due to a thrombotic or embolic occlusion of one or more supplying arteries. The normally high oxygen requirement of neural tissue makes the brain more susceptible to decreased perfusion and ischemic injury [2]. Injury of the cerebral parenchyma begins within minutes after vessel occlusion [3]. Cellular death continues for hours, days, and even weeks following the ischemic event [3]. Cytokines released from the ischemic parenchyma as well as damaged vasculature initially promote an acute inflammatory response followed by a chronic inflammation in the hours and days following vascular occlusion eventually amplifying the initial ischemic damage [2]. Despite the initially detrimental immune response, proinflammatory factors released from the ischemic tissue and the inflammatory response ultimately promote neurogenesis in neurogenic areas of the brain and lead to neural progenitor migration to the ischemic penumbra [4, 5]. Upregulated neurogenesis and neuronal progenitor migration have been repeatedly observed in the rodent and human subventricular zone (SVZ) following cerebral ischemia [4-6]. However, this putative endogenous repair system is not sufficient to repair all stroke damage. The extent and effectiveness of this endogenous repair mechanism in the weeks and months following injury remain under investigation. With the very limited stroke treatments available and a very small percentage of patients actually benefiting from the only available drug for ischemic stroke, tissue plasminogen activator (tPA) much of the research effort in the field is now dedicated to finding stroke therapeutics that enhance stroke repair.

This chapter focuses on how chemotactic factors and chemokines play definitive roles in neural progenitor recruitment, proliferation, migration, and limited tissue replacement. Chemokines play a role in regulating migration to the site of neuroinflammation. Adult neural progenitor cells widely express receptors for chemokines. For example, stromal-cell-derived factor-1 (SDF-1) is a major chemotactic factor that is upregulated in the infarct and peri-infarct areas after stroke [6, 7]. SDF-1 and its receptor, CXCR4, are involved in progenitor homing from the SVZ to the infarct, neuronal differentiation, and cell survival. In chamber assays, neural progenitors migrated in a SDF-1 dose-dependent manner [7]. It has been demonstrated that neural progenitor cells migrate further to the infarct region after ischemic stroke compared to controls that receive no stroke [7]. When SDF-1 is neutralized with anti-SDF-1 antibody, migration is attenuated and the stroke outcome is worse [7]. Several other trophic factors are involved in neural progenitor migration and proliferation including BDNF and VEGF.

In addition to chemokines which are indispensible for progenitor migration, the neuro-vascular unit provides physical scaffolding for migrating cells [8]. The pattern and timing of migration and the microenvironment of the neurovascular unit allow migration to be successfully targeted to the injury area. To study the interplay of chemoattraction and the neurovascular microenvironment, researchers have employed various models of stroke and migration as well as several established and novel techniques for visualizing migration in the ischemic brain.

Several stroke treatments have been explored for the enhancement of progenitor migration. For example, peripheral sensory stimulation and enriched environment are modes of enrichment that have enhanced progenitor recruitment to the infarct area. Together, understanding the mechanisms of chemoattractant functions and the microenvironment of the injury area are necessary to promote tissue repair after ischemic injury and ultimately treat stroke patients.

II. Ischemic Stroke Damage and Cell Death Mechanism

The brain has a significantly greater vulnerability for decreased perfusion than any other organ due to the intrinsically high metabolic rate required to maintain homeostasis within the cerebral parenchyma [3]. Tissues receiving less than 20% of normal blood flow undergo anoxic depolarization, energy failure, and cell death following vessel occlusion. Ischemic damage resulting from stroke is heterogeneous. Cell death from ischemic insult occurs in acute and chronic phases. The ischemic center area damaged the severe ischemia is referred to as the infarct core [9]. In the minutes and hours following vascular occlusion, the largely decreased perfusion and metabolic failure cause cell membrane and mitochondrial membrane depolarization, subsequently leads to glutamate-mediated excitotoxicity as a result of uncontrolled Ca^{2+} influx, intracellular acidosis, disruption of ionic homeostasis, and release of free radical species [3, 10]. The neurovasculature disruption leads to tissue edema and blood brain barrier (BBB) degradation further increasing cellular mortality. Most cells that die at this time are necrotic in nature and make up the ischemic core within a few days. Parenchyma adjacent to the ischemic core is termed the peri-infarct zone, or ischemic penumbra [9, 11]. Cells in the penumbra are potentially salvageable with neuroprotective interventions. The penumbra receives collateral perfusion to temporarily maintain cellular viability [9, 11]. The injurious factors released from the ischemic core region and the partially-reduced blood flow have significant impacts on this region. The stressful environment may activate the program for controlled cell death, or apoptosis in a delayed fashion (hours and days after ischemic insult). Due to the continuous expansion of necrosis and activation of apoptotic cascades, cells may die in the penumbra from necrosis or apoptosis. However, recent evidence suggests that many cells actually die from a mixed form of both necrosis and apoptosis [12]. In addition, autophagy may contribute to the ischemic injury during the sub-acute phase of cerebral ischemia [13]. Comprehensive reviews can be found for detailed overview of the heterogeneous nature and cellular/molecular mechanism of stroke damage [14]. This review will focus on the mechanism of post-ischemia regeneration specifically related to endogenous neurogenesis and cell migration.

III. The Role of the Inflammatory Process in Neural Progenitor Recruitment

A. Neurogenesis and Rostral Migratory Stream Migration

Regenerative activities in the adult brain was observed at least 30 years ago [15]. Not until recently has it been widely accepted that endogenous neurogenesis occurs in adult mammalian brain which contradicts the long-standing dogma about the brain's inability to regenerate [16]. Compelling evidence has demonstrated that there is a continuous production of neurons in adult non-human primates as well as in humans [17, 18]. Increasing evidence supports that specific regions of the adult brain such as the SVZ and the subgranular zone (SGZ) are capable of responding to injurious signals from other areas of the brain by increasing the proliferation of neuroblasts [5, 19]. This proliferative response also occurs in response to non-injurious stimuli such as in the SGZ after exercise [20, 21]. This response and the migration of newly formed neural progenitor cells towards the injured tissue are regarded as parts of repair process in the adult brain. Under normal physiological conditions, adult mammalian brains have endogenous stem cells that migrate in a network of chains along the rostral migratory stream (RMS) to the olfactory bulb where they differentiate into interneurons and are functionally incorporated into existing neural circuitry [22, 23]. These cells also migrate from the subependymal zone of the lateral ventricle into the olfactory bulb in rats [24]. The reasons for the renewal of these olfactory bulb cells are not clear. This migratory stream can be also re-directed to different areas and help in repairing damaged brain tissue. The environmental cues that can affect the behavior of progenitor migration are physiological processes that can be applied to the translational area of research. So far, numerous factors have been identified of being capable of influencing neural progenitor proliferation and neuroblast migration after a cerebral ischemic insult [25].

B. Inflammation and Neural Progenitor Migration

Even though inflammation is a significant cause of cell death in stroke, inflammation can paradoxically play a role in neurogenesis, migration and regeneration. Proinflammatory factors such as tumor necrosis factor-α (TNF-α) can increase the proliferation and recruitment of endogenous neural progenitors to the ischemic infarct [26]. These factors can also enhance progenitor survival, and induce the migration of neuroblasts from the SVZ towards the site of ischemic injury [4, 5, 26]. Furthermore, enzymes involved in the metabolism of TNF-α, namely TNF-α-converting enzyme TACE, are upregulated in the SVZ of ischemic brains an4d play major roles in stroke-induced SVZ proliferation [27]. Accordingly, numerous chemotactic agents upregulated by ischemia and inflammation can be classified as both pathologic and therapeutic depending on the time course and the extent of their expression [28]. Matrix Metalloproteinase (MMP) upregulation, for example, is highly detrimental in the acute phase of stroke, however, MMPs are also upregulated by cells migrating from the SVZ to the ischemic penumbra for repair [29]. MMPs in the chronic phase of stroke are involved in regeneration and tissue repair of the damage [29]. Blocking MMPs in the weeks following injury significantly decreases neural progenitor migration and is detrimental to functional recovery [30, 31]. Additionally, factors such as monocyte chemoattractant protein-1 (MCP-1) secreted by astrocytes and microglia not only serve as chemoattractants for lymphocytes, basophiles and macrophages, but also neural progenitors [32]. Knocking out MCP-1 expression significantly decreases neuronal progenitor migration *in vivo* [33].

C. Post-Injury Chemotaxis and Migratory Pathways

In the days and weeks following stroke, SVZ neural progenitor proliferation is greatly enhanced. The migrating progenitor cells are diverted from the RMS to the ischemic infarct [34]. Neuronal progenitor migration begins within hours of injury, peaks at approximately 3 weeks, and continues for several months following the focal ischemic event [34-36]. Diffusible chemotactic factors promote neural progenitor migration in a dose-dependent manner as the extent of ischemic injury has been demonstrated to determine the extent of subsequent neurogenesis and neural progenitor migration [7, 37]. However, current evidence suggests that the spontaneous post-ischemia regenerative activity cannot fully replace stroke-damaged tissue and its disrupted circuitry. Therefore, promoting and optimizing neurogenesis has been the focus of regenerative research.

Neural progenitor migratory pathways to the ischemic penumbra tend to be closely associated with cerebral vasculature where vessels provide a scaffold for migration [8, 38]. Several chemotactic factors released from the ischemic penumbra are pleiotrophic and promote neurogenesis as well as angiogenesis [25]. Erythropoietin (EPO) and vascular endothelial growth factor (VEGF) for example are commonly known to enhance angiogenesis, but also upregulate neurogenesis as well as direct neural progenitor migration [39]. Another versatile chemotactic factor is SDF-1α which is upregulated by activated astrocytes and vascular endothelial cells during ischemic injury [4, 7, 40, 41]. In addition to promoting neurogenesis in the SVZ, and guiding neural progenitors to the penumbra via chemotaxis, SDF-1α promotes angiogenesis [4]. Moreover, blocking CXCR4, the receptor for SDF-1α, decreases neural progenitor migration, suggesting that this chemotactic factor plays a significant role in guiding post-injury neural progenitor migration [4, 42, 43].

IV. Stroke and Cell Migration

As part of the response to an ischemic insult, the surviving cells in the ischemic core and penumbra release several trophic factors that play roles in repair and tissue regeneration. Many of these factors are involved in the recruitment of neural progenitor cells to the damaged region for neural and vascular repair. We will focus on the roles of chemotactic factors, proteases such as MMP and cellular components involved in adhesion such as β integrins in neural progenitor migration.

A. Chemotactic Factors

Chemotactic factors and chemokines play imperative roles in neural progenitor recruitment and ultimately tissue replacement. Chemokines play a role in regulating migration to the site of neuroinflammation. After ischemic stroke, there is a local production of chemotactic trophic factors. These include, but not limited to SDF-1, VEGF, and brain-derived neurotrophic factor (BDNF). With the increased levels of local attractive cues, neural progenitor cells migrate further after ischemic stroke compared to controls that receive no stroke and many of them can move toward the ischemic cortex region. Adult neural progenitor cells widely express receptors for chemokines [6, 44]. It has been noticed that even with the guidance of these factors, endogenous progenitor cell integration and replacement is not sufficient to repair stroke damage.

B. Cytokines and Inflammatory Factors on Progenitor Recruitment

New studies have revealed that the role of inflammation on neurogenesis is much more complex than what was previously thought. Inflammatory reactions especially from microglia can be beneficial after injury. Some aspects of inflammation can support neurogenesis, differentiation, proliferation, migration, and survival [45]. Microglial activation may be both helpful and harmful depending on the balance of the factors secreted [45]. Recent studies have shown that neuroinflammation and microglia can have beneficial effect on neurogenesis [45, 46]. In a co-culture system of neural stem cells and microglia or growing stem cells in microglia-conditioned media, a protective effect on stem cells was seen; neural stem cells in conditioned media survived longer than those of control cultures [47].

Inflammation can provide guidance cues. Inflammatory cues direct the migration of potential repair progenitors. Both transplanted and endogenous neural progenitors have been shown to migrate to the site of neuroinflammation and chemokines. For example, microglia are immunoreactive cells that are activated and concentrated in areas of injury. Microglia are a source of cytokines like IL-6, TNFα, and IL-10 and also a source of neurotrophins like BDNF, and other chemokines [48, 49]. In a study where conditioned media from microglia cultures was applied to precursor cells and assayed for its effects on migration and differentiation, increase in migration with microglia-conditioned media was seen in a Transwell migration assay [47]. There was also an increase of migration in cells treated with lipopolysaccharide (LPS)-stimulated cell conditioned media compared to non-stimulated cultures. Interestingly, this study also showed that microglia-conditioned media increases the differentiation of precursors to neurons but not astrocytes [47]. While microglia may have a neurotoxic and neuroprotective role, this work supports the hypothesis that microglia also play a role in neuroprotection. When an inflammatory stimulus was injected into a hippocampal slice culture, eGFP-labeled neural progenitors transplanted in the slice migrated towards the site of the inflammatory stimulus 7 days after injection [26]. The injection consisted of TNF-α and interferon-γ, the bacterial toxin, LPS, the human immunodeficiency virus-1 coat protein glycoprotein 120, or a β-amyloid-expressing adenovirus. This suggests that cytokines from sites of neuroinflammation can attract neural progenitors towards the inflammatory gradient [26]. The ischemic lesion is thus a reactive area with inflammation, cell death and also several chemoattractants upregulated to assist in repair.

C. SDF-1 is a Chemoattractive Factor

Chemokines have been mainly studied in the context of inflammation. For example, they are greatly involved in neutrophil chemotaxis during the inflammatory process. It is clear now that directed cell movement in developmental or pathological processes relies heavily on chemokines. Chemokines are chemotactic cytokines that are expressed both in peripheral system and the CNS. They recruit embryonic, adult progenitor, and cancer cells. SDF-1 is a constitutively expressed chemokine involved in CNS processes via its G-coupled protein transmembrane receptor, CXCR4 [50, 51]. In the peripheral system, SDF-1 is highly chemotactic for lymphocytes and it directs the migration of hematopoietic cells from the fetal liver to the developing bone marrow [52, 53]. SDF-1 is a leukocyte localizing cue for leukocyte entry into the CNS parenchyma [54]. Since the CXCR4 receptor may have other functions besides in inflammation and neutrophil chemotaxis [55], the role of CXCR4 in areas not related to immunology has been studied. For example, SDF-1 knockout mice have impaired cerebellar neuronal migration where the external granule cell layer develops irregularly [53]. Similar abnormalities such as impaired myelopoiesis, cardiogenesis, and β-lymphopoiesis in both SDF-1 and CXCR4-deficient mice illustrate that SDF-1 and its receptor are involved in other functions aside from the immune system [53, 56].

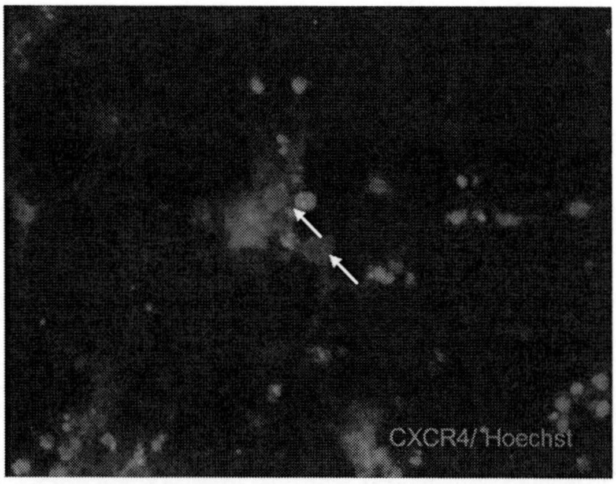

Figure 1. CXCR4 receptors are expressed on embryonic cortical neurons. White arrows indicate positive staining of cytoplasmic CXCR4. Antibody: R&D MAB21651.

Like many chemokines, SDF-1 plays a role in cell trafficking and is a potent chemoattractant to many cell types. CXCR4 is expressed on many progenitor cell types including hematopoietic stem cells [57], skeletal muscle satellite progenitor cells [58], neural progenitors [44], hepatic oval cells [59], as well as retinal pigment epithelium progenitors [60]. CXCR4 is also expressed in cortical neurons (Fig. 1). The SDF-1-CXCR4 signaling plays a role in trafficking organ progenitor cells during development for embryogenesis, organogenesis, and tissue and organ regeneration in adulthood. SDF-1 and CXCR4 expression can be found in the stem cells derived from the inner cell mass of the blastocyst and SDF-1 increases the survival and migration of these cells [61]. To illustrate the chemoattractive abilities of SDF-1, meninges from SDF-1-deficient mice were transplanted onto host mice and found to exhibit no attraction of neural progenitors to the transplant whereas wild-type animal meningial transplants exhibited progenitor migration in the host [62]. CXCR4 is found on both embryonic and adult stem cells and it directs migration of cells under normal developmental and pathological circumstances. When the subventricular neural stem cells were injected with a CXCR4 siRNA retroviral vector, migration of these progenitors was attenuated compared to the empty vector [63].

Pathologically, since SDF-1 is a potent chemoattractant, it also plays a role in cancer and tumor metastasis because cancer cells express CXCR4 on their surface. SDF-1 has been also shown to be involved in tumor formation [64]. In all, SDF-1 plays a role in cell trafficking in several organs and body systems for physiologic processes during development and adulthood as well as in pathological processes. We will discuss its role as a progenitor chemoattactant in the ischemic brain.

D. SDF-1 and Ischemia

We and others have shown that SDF-1 is upregulated in the infarct and peri-infarct area in a gradient-fashion [7, 65, 66]. Its expression peaks at 7 to 14 days after the insult [67-69]. SDF-1 is a pleiotrophic factor that can benefit the brain after ischemic injury. SDF-1 increases cell survival through increasing the expression of anti-apoptotic proteins. Its neuroprotective protective effect has been demonstrated *in vitro* and *in vivo*. Cells exposed to hydrogen peroxide (H_2O_2) exhibited fewer caspase-3-positive cells and more Bcl-2 expression when SDF-1 was added [70]. Similarly, when SDF-1 protein was injected into the ischemic brain, infarct size decreased and motor function increased demonstrating that SDF-1 contributes to neuroprotection and repair in the brain [70].

SDF-1 is produced and released by astrocytes and endothelial cells from the tissues surrounding the ischemic injury [67]. SDF-1 works through binding its membrane-spanning G-protein coupled receptor to initiate divergent cascades leading to chemoattraction, survival, cell proliferation, and gene transcription [71]. Upon ligand-receptor binding, the Gα, Gβ and Gγ subunits that are normally associated under basal conditions dissociate. Downstream PI3K can activate AKT which can lead to cell survival. Cell survival may also be promoted through p38 and ERK as well [72]. This pathway, however, could lead to tumor cell survival as well.

The intracellular cascades for SDF-1 signaling are currently being elucidated. Intracellularly, the transcription factor, hypoxia-inducible factor-1 (HIF-1) has been proposed to be a key player in the induction of SDF-1. Many adaptations occur at the cellular level in response to hypoxia and HIF-1 is the master regulator of hypoxia in the cell. HIF-1 regulates SDF-1 gene transcription in endothelial cells demonstrating that SDF-1 is selectively upregulated in areas of decreased oxygen tension [73]. Consequently, HIF-1 expression increased SDF-1-induced adhesion and migration of CXCR4-positive progenitors to the ischemic injury. When blocking SDF-1 expression or neutralizing CXCR4 on migrating cells, progenitor migration was attenuated. SDF-1 is highly expressed in the ischemic tissue especially in the vasculature including the endothelial cells and the vascular lumen [73]. In human endothelial cells *in vitro* SDF-1 secretion after ischemic conditions was examined. After exposing cells to 1% oxygen, there was a nine-fold increase in SDF-1 secretion at 12 hours [73]. Cleverly, in the same study, using cells transfected with a mutated HIF-1 binding site, an attenuation of hypoxia-related gene expression including SDF-1 was observed. These experiments demonstrate the close interplay of HIF-1 and SDF-1 under ischemic conditions.

E. SDF-1 as a Potential Therapeutic: Bone Marrow and Embryonic Stem Cell Chemoattraction

SDF-1 not only plays a role in the homing of endogenous progenitors from the SVZ, but can also home bone marrow stem cells and transplanted cells to the ischemic infarct. Endogenously, SDF-1 can mobilize hematopoietic cells from the bone marrow to the peripheral blood and then to the stroke infarct [53, 70]. $CD34^+$ cells express CXCR4. CXCR4 and SDF-1 draw hematopoietic stem cells from the bone marrow to the peripheral blood [74]. Rats injected with SDF-1 30 minutes after stroke revealed more 5-bromodeoxyuridine (BrdU)

-labeled engrafted cells in the ipsilateral cortex 3 days after stroke compared to the controls [70]. Similarly, GFP-positive bone marrow cells from a GFP transgenic rat injected into a donor had greater GFP cells in the penumbra after SDF-1 local injection compared to the control [70]. With SDF-1 upregulation in the penumbra, neural stem cells transplanted into the contralateral hemisphere also migrate towards the site of ischemia [4]. These results suggest that understanding the characteristics of SDF-1 and its chemoattractive abilities for endogenous and exogenous stem cells provides a molecule target for a potential therapy for ischemic stroke.

V. MULTIPLE PROTECTIVE AND REGENERATIVE ROLES OF NEUROTROPHIC FACTORS

Five members comprise the VEGF family in the mammals: VEGF-A, -B, -C, -D and placental growth factor [75, 76]. The ligand, VEGF binds to VEGF receptors (VEGFR) which are tyrosine kinases that phosphorylates downstream molecules. VEGF, first discovered as an angiogenic factor and vascular permeability factor, is now understood to be pleiotrophic and play several roles in neurogenesis and neuronal survival [77, 78]. It is an important regulator of angiogenesis due to its role in endothelial cell proliferation and progenitor migration [79].

A. Neurogenic Effects of VEGF

Research has demonstrated that VEGF has both proliferative and neuroprotective actions. *In vivo*, intracerebroventricular (ICV) infusion of VEGF increases progenitor proliferation in the dentate gyrus (DG) of the hippocampus and the SVZ [80]. In transgenic VEGF-overexpressing mice without stroke, basal levels of BrdU-labeled cells were approximately 20% greater compared to wild-type mice [81]. At the maximal survival concentration of VEGF, treatment doubled the number of surviving cells in *in vitro* models of ischemia [82].

Many groups have focused on the effect of VEGF and its effect on angiogenesis and functional recovery after ischemic stroke [25, 39, 83-85]. Its role in angiogenesis has been well studied, but VEGF is a pleiotrophic factor that also plays roles in chemoattraction and neurogenesis. It also plays a pivotal role in developmental and adult neural progenitor migration through its action on VEGFR-2 [86]. Hypoxia upregulates VEGFR-2 [87] and VEGF stimulates neurogenesis in the SVZ and the SGZ of the hippocampal DG through VEGFR-2/Flt-1 pathways [85, 88]. For example, under hypoxia and glucose deprivation, an *in vitro* model of cerebral ischemia, VEGF reduced cell death in primary neuronal cell cultures and HN33 cells, an immortalized hippocampal neuronal cell line. In this study, neurogenesis was shown due to VEGF's tyrosine phosphorylation action on the VEGFR-2 but not VEGFR-1 since the HN33 cell line does not express VEGFR-1 receptors [82]. When phosphatidylinositol 3-kinase (PI3K) was inhibited, the protective effect of VEGF diminished suggesting that PI3K/Akt plays a downstream role in VEGF-mediated neurogenesis [82].

B. VEGF in Stroke and Migration

VEGF expression increases after an ischemic insult. It is upregulated as a gradient in the infarct area. VEGF and VEGR-1 mRNA expression increases as early as 2 hours after stroke and VEGFR-2 increases 4 hours after stroke [89]. Transgenic mice constitutively expressing human VEGF in neurons have increased neurogenesis in the SVZ and increased chain migration to the ischemic infarct [81]. The infarct volume is reduced in mice overexpressing VEGF [81]. VEGF overexpression increases neurogenesis in the brain after ischemia [90]. *In vitro* studies suggested that progenitor migration requires a "priming" of SVZ-derived progenitors with FGF-2 for migration towards VEGF beads [86]. Pharmacologically upregulating VEGF and other trophic factors in the penumbra has shown to be beneficial. Recent work with Atorvastatin, a drug that reduces blood cholesterol levels and increases VEGF, VEGFR-2, and BDNF in the ischemic penumbra has shown beneficial improvement in brain plasticity, stroke outcomes and functional recovery [25]. Atorvastatin also increased migration of SVZ cells *in vitro* and *in vivo*, synaptophysin-positive cells, and angiogenesis. According to these studies, VEGF, VEGFR-2, and BDNF are most likely to contribute to these processes.

On the contrary, despite its beneficial effects, VEGF is well-known to induce blood brain barrier leakage acutely after stroke. Thus, current strategy of VEGF therapy is: 1) an acute inhibition of VEGF early after ischemia to reduce BBB leakage and the risk of hemorrhage and 2) delayed increases in VEGF signaling to improve migration and functional recovery after stroke [91].

C. BDNF in Stroke and Migration

Similar to SDF-1 and VEGF, BDNF has many neurotrophic effects. *In vivo*, BDNF increases the survival of newly-born neurons [92]. BDNF increases in newborn neurons [93, 94]. However, currently there is some controversy over whether or not BDNF contributes to SVZ neurogenesis [95]. Its receptor, TrkB is a tyrosine kinase cell-surface receptor. In ischemia, BDNF plays a role in preventing ischemic cell death by binding to its receptor and activating intracellular protein kinase B (i.e. Akt), mitogen-activated protein (MAP) kinases and extracellular signal-related kinases (ERKs) [96]. BDNF has been known to differentiate and help in the survival of cells [97]. It has also shown to be involved in reducing apoptosis after ischemia and reducing infarct size. It decreases the expression of proapoptotic proteins like Bax, and increases the expression of antiapoptotic proteins like Bcl-2 [98]. Ischemia causes an increase in mRNA expression of BDNF ipsilateral to the stroke lesion [99]. Systemically administered BDNF (intravenously administration) in ischemic stroke animal models has shown to improve functional improvement in the Rota-rod, adhesive removal tests and the neurological severity score (NSS) 6 weeks after stroke compared to vehicle-treated animals [96].

BDNF plays a role in cerebellar and cortical interneuron migration during development [100, 101]. The aforementioned work with Atorvastatin suggests that BDNF, VEGF and VEGFR-2 play a role in plasticity and an increase in migrating cells [25]. Precursor cells migrating along the rostral migratory stream to the olfactory bulbs utilize BDNF and its receptor, TrkB. Their expression is found throughout the migratory tract *in vivo* [102]. In

SVZ tissue culture explants that were treated with BDNF, there was a 2.5-fold increase in migration when compared to control [102]. Explants that were treated with the TrKB inhibitor, K252a blocked BDNF-induced migration. In ischemic BDNF-treated animals, hippocampal neurogenesis increased as visualized with BrdU labeling in the hippocampus [102]. Animals injected with an adenovirus overexpressing BDNF in their lateral ventricles exhibited new neurons in the rostral migratory stream and subsequently new neurons in the olfactory bulb. There was an increase in BrdU-positive cells in the neostriatum and RMS that were both NeuN- and βIII-tubulin-positive cells. The BrdU-positive cells in the neostriatum expressed calbindin-D28K, a marker for medium spiny neurons. Many of these cells expressed GAD67, a marker for GABAergic neurons. In the adult forebrain, intraventricular injection of BDNF increased the number of mature neurons in striatum and septum, as well as the thalamus and hypothalamus [103]. BDNF overexpression in the adult ventricular zone increases the recruitment of new neurons to neurogenic and non-neurogenic areas of the brain [104]. Adulthood neurogenesis in these areas are novel observations.

VI. Adhesion Molecules and Migration

A. β1 Integrins Play a Role in Progenitor Division and Migration after Stroke

Integrins are cell surface receptors that play key roles in cell adhesion and migration. Under normal developmental events, plated neurospheres exhibit chain-like migration *in vitro* [105]. Unlike radial migration which utilizes a glial fiber scaffold [106], neural progenitors of the tangential rostral migratory stream have been shown to migrate without the need for a glial scaffold [107]. When certain integrin subunits are blocked, slower and/or fewer migration occurs. For example, blocking α6β1 integrin reduces chain migration [105]. This suggests integrin involvement in chain migration of SVZ cells during repair.

VII. Characteristics of Progenitor Migration

A. Progenitor Migration Requires a Physical Scaffold

Migrating progenitors require a physical scaffold to be able to physically move from the SVZ to the ischemic site in the brain. In ischemic brains, neuroblasts migrate in chains or individually to the ischemic infarct [108]. It has been observed that DCX-labeled neuroblasts originating from the SVZ laterally migrate to the striatal infarct. The chain formed at the border of the SVZ and extended towards the striatum. With time-lapse microscopy, researchers tracked the cells and showed they migrated at a rate of 28.67 ± 1.04 μm/h [108]. Interestingly, the neuroblasts would extend and retract their processes probing the microenvironment within the ischemic tissue and eventually extending many processes when encountering a vessel. This suggests that the cells probe their microenvironment to guide them to the infarct [108]. The migrating DCX-positive cells were closely associated with cerebral vessels, clustered around them. Some migrating chains were seen to move in parallel to the vessels. About half of cells migrating to the adult olfactory bulb were observed to use blood vessels as a scaffold for migration [109]. These migrating cells interact with the end-feet of astrocytes and their extracellular matrix in the olfactory bulb. In stroke, cells migrate in association with reactive astrocytes and vessels [110]. They migrate along astrocytes and blood vessels.

In the SVZ and SGZ of the normal unaffected adult brain, progenitors associated with endothelial vessel cells. This interaction is termed the neurovascular niche [111]. This is a link between neurogenesis and angiogenesis. After ischemic insult, the vessels release both SDF-1 and Angiopoietin 1 (Ang-1) which are involved in neurogenesis and vasculogenesis. SDF-1 and Ang-1 induce neuroblast migration and function behavioral improvement. The molecular expression of microvessels around the infarct changes with ischemia [112] supporting that a neurovascular niche is created around the infarct facilitating migration to the stroke area along blood vessels. PSA-NCAM-positive precursor cells were observed to be in close proximity and associated with endothelial cells in the peri-infarct area [8]. 7 days after stroke, they found that newly born endothelial cells were closely associated with BrdU/ DCX-positive cells. Similarly, others have found that 35% of new progenitors born after stroke were closely associated with a vessel (within 5 microns) [110]. Thus, the neurovascular niche is integral to the migratory repair process.

B. Timeline of Progenitor Migration and the Fate of Progenitors

Migration of newly born neurons as well as neuroblasts formed before the stroke can be seen in the stroke infarct about 2 to 4 weeks after ischemia [36]. The peak number of cells in the SVZ is approximately 3 weeks after stroke and drops to its lowest point around 8 weeks [36]. By tracking BrdU-positive cells, the migration of progenitors and dividing cells timeline follows a similar pattern to neurogenesis in which BrdU-labeled cell migration peaks around 3 weeks after ischemia and reaches a nadir around 6 weeks after stroke [36]. These cells go on to differentiate and express markers of mature neurons including region-specific labels such as calbindin and dopamine and cAMP-regulated phosphoprotein-32 [5]. Stroke damage to the striatum exhibits progenitors that migrate to the striatum and differentiate into striatal medium-sized spiny neurons [5, 110]. In our group, we were able to demonstrate that 30 days after stroke, a fraction BrdU-positive cells exhibited the mature neuronal marker, NeuN in the penumbra [65].

VIII. Enriched Environment on Migration

An enriched environment can provide nourishment and trophic support for repair of the ischemic brain. For example, focal stroke models with an MCA occlusion targeting the whisker barrel cortex lends itself to functional stimulation studies. A modular organization of cortical capillary vessels is found resembling the whisker barrel cortex. When row C whiskers were stimulated, a corresponding increased blood flow can be found in arterioles on the contralateral cortex in row C barrels [113]. This suggests that there is a functional correlation between whisker stimulation and vascular supply. Moreover, stimulating the whiskers increase in immediate early response gene products such c-fos expression in the contralateral cortex. Applying rostro-caudal whisker stimulation three times a day to the contralateral side of the whisker barrel stroke hemisphere enhances angiogenesis [85, 114]. Enriched environment has been shown to increase trophic support and enhance the migration of progenitors. With enrichment such as whisker stimulation, we demonstrated that whisker stimulation after a stroke in the barrel cortex increased endogenous neurogenesis and progenitor migration to the infarct [65, 85]. Whisker stimulation has also been shown to increase trophic factors such as nerve growth factor (NGF) and IL-1β [115]. Similarly, exogenous SVZ stem cell transplants exhibit a similar behavior where upon whisker stimulation, the cells display enhanced migration towards the stroke [116]. In our work, more endogenous migrating cells differentiated into neurons with whisker stimulation compared to stroke only animals [65]. Thus, the barrel cortex stroke model and whisker stimulation method can serve as a model of physical rehabilitation and repair of stroke-disrupted brain structures.

IX. Techniques to Study Post-Ischemia Cellular Migration

To study the interplay of chemoattraction and the neurovascular microenvironement, researchers have employed various models of stroke and migration as well as several established and novel techniques for visualizing migration in the ischemic brain. These include GFP virus labeling of neural progenitors, and MRI tracking of cells.

A. Models of Ischemic Injury and Migration

To better understand the various events of stroke development, the migration of progenitor cells following injury, and to develop potential stroke therapeutics, several models of ischemic injury and progenitor migration are commonly tested. Current models employ *in vitro* neural progenitor cultures, mixed neuronal and glial cultures, brain slice cultures, as well as several in *vivo* ischemic stroke models [117, 118].

a. In Vitro Modeling of Hypoxia Insults and Cell Migration

In vitro models of hypoxia insult and glucose-oxygen deprivation (OGD) insult mimic more or less the ischemic condition in the brain. Post-insult cell proliferation and migration tests allow for isolated analysis of individual biochemical processes without systemic influences under highly controlled manipulation of individual variables as well as real time measurement of the cellular responses to manipulation [119, 120]. *In vitro* models of cell migration include the chamber assay, wound healing assay, and brain slice cultures [121, 122]. However, *in vitro* modeling of ischemia and migration does not accurately represent what happens *in vivo* and remains preliminary until confirmed by animal models.

The wound healing assay, also referred to as the scratch test involves scratching cell cultures with a pipette tip to visualize cell movement into the scratched empty "wounded" area (Fig. 2) [123]. This is a simple, straightforward test where one can observe time-dependent cell morphological changes and real time movement. The migration direction and distance can be easily quantified. Time-lapse movies can be made to demonstrate the migration activities.

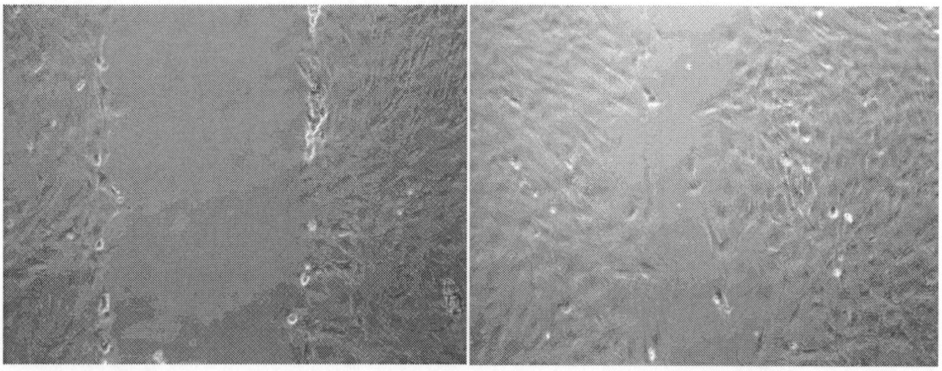

Figure 2. The scratch test is a simple assay to assess cell migration *in vitro*. A clean pipette is used to scratch and clear an area for cell invasion.

The chamber assay, also known as a Transwell assay, Boyden chamber, or invasion assay is another way to assess cell migration *in vitro*. It involves culturing cells in a migration chamber mimicking a 3-dimentional environment. Cells can be exposed to differentially chemotactic agents to quantitatively analyze cellular migration [124]. A chemoattractant is added to the bottom chamber to attract progenitor cells from the upper chamber (transwell insert) into a membrane. Cell invasion is then quantified by fixing and staining the membrane to visualize the number of cells that have penetrated the membrane. For example, study from our group has shown that FAK activation associated with an upregulation of the Kv2.1 channel significantly increased directed migration of BMSCs in the Transwell assay [125]. The assay can be used to show a dose-dependent migration of neural progenitor cells in responding to a chemoattractive cue [42]. These types of *in vitro* studies are important for identifying novel molecules that can regulate cell migration under normal as well as pathological conditions.

b. Slice Cultures Models

Originally developed in 1981 by Gahwiler, brain slice culture is considered to be the best *in vitro* model to represent *in vivo* stroke pathophysiology [122, 126, 127]. The two general categories of brain slice cultures, acute and organotypic are used to study neurogenesis and cellular migration [121, 127]. Acute brain slice cultures tend to be short-lived (for short-term studies) and have been used to quantify the mechanisms of cell death [127]. Organotypic brain slice cultures typically have an extended longevity and are used to quantify cell proliferation and migration [127, 128]. Ischemic stroke model applications of the organotypic brain slice has been used to observe migration speed from the SVZ towards the ischemic striatum [108]. Ischemic stroke is induced first in rodents and sliced sections of the brain can be made for examination of neural progenitor migration. The DCX-labeled eGFP-cells can be followed for formation of migration chains extending laterally towards the injury region [108]. Another example of using this technique is to study the effect of ischemic lesions on the motility of progenitor cells. Cell migration on the stroke brain slices showed greater migration compared to control brain slices with no stroke [7]. Thus, brain slice cultures can provide a model that allows studying an isolated system while retaining many features of actual *in vivo* model.

c. In Vivo Modeling of Ischemic Stroke and Cellular Migration

Damage from ischemic stroke tends to be confined to the neural parenchyma and neurovasculature, however the underlying pathophysiology is mediated substantially through interactions between the central nervous system and systemic factors such as components from the immune response. *In vivo* models have been developed to replicate both global and focal ischemia to provide a representative multi-system model of human stroke pathophysiology as well as to test the potential effectiveness of various therapeutics. *In vivo* stroke injury models allow for the study of migration of the whole range of stroke injuries and for studying endogenous progenitor migration to the infarct.

Due to the many manifestations of stroke injury depending on the occlusion location, duration and cause, we have used different types of models to study stroke pathophysiology and neural progenitor migration since the location of the blockage, duration of ischemia, and the infarct size vary in patients [129]. Other variables to be modeled include differences in age, sex, lifestyle, and preexisting conditions. In is necessary to consider all these factors when modeling ischemia and testing stroke therapies. For example, animal models of diabetes and hypertension have been studied in the context of stroke injury to delineate the intricate relationship between the underlying pre-existing conditions and stroke outcomes [130, 131].

Animals most commonly used to model ischemic stroke and post-injury neural migration are rodents, however larger mammals such as cats, pigs, and non-human primates are also invaluable for translational research [117, 118, 132]. There are benefits and disadvantages to the choice of animal model used. Rodents are inexpensive, easily maintained, easy to handle and, behaviorally testable. Mice are commonly used because of the availability of a huge number of genetic models for genes that are involved in neuronal development and migration, immune regulation and neuronal function and viability. Another benefit is the smaller brain size of the mouse allowing for easier tissue handling and histological analysis as well as the ability to examine the whole brain without additional cost [133]. However, surgical manipulation for mice requires finer skills. Also, comparatively they do not have as much white matter or cortical area as other mammalian models do. Unlike humans, rodents have lissencephalic, or an unconvoluted brain and less cortical area. Genetically modified rats are less available. However, rats may have the advantages of a larger brain/cortex than mice and they are more similar to humans in their cranial circulation [133]. Like mice, rats can be behaviorally and functionally evaluated as well.

Rabbits and pig models have more white matter and cortex. Many non-rodent models are similar to human in that their brains are gyrencephalic, or convoluted133. Pigs have a more similar vasculature to humans compared to other animals. However, it is not readily feasible to genetically manipulate pigs and rabbits. Non-human primates are the closest models to humans, but they are expensive to maintain, handle and are not as accessible as the other animals. These non-rodent larger stroke models lend themselves to complex physiological monitoring like glucose levels, blood gas values, and direct blood pressure measurements that is not possible in rodents but more closely resembling the human situation [133].

Table 1.

Category		Model	Notes	References
In Vitro	Migration	Boyden Chamber Assay	Also known as the chamber, Transwell or invasion assay, this technique involves adding cultured cells or slices of neural tissue to a migration chamber to test there response to chemotactic agents.	Klemke, R. L. et al. CAS/Crk coupling serves as a "molecular switch" for induction of cell migration. J Cell Biol 140, 961-972 (1998).
		Wound Healing Assay	Also referred to as the scratch test, this technique involves scratching cultured neurons with a pipette tip in order to determine cellular migration and proliferation following induced injury.	Sieg, D. J., Hauck, C. R. & Schlaepfer, D. D. Required role of focal adhesion kinase (FAK) for integrin-stimulated cell migration. J Cell Sci 112 (Pt 16), 2677-2691 (1999).
	Stroke	Oxygen glucose deprivation	OGD is she most widely used in vivo technique for simulating ischemic stroke pathology in vitro. The technique involves the suspension of cells in a glucose free medium within a hypoxic environment. Variations of this technique have been used on primary neuronal cultures, mixed neural cultures, and slice cultures, to examine the cellular response to ischemic conditions.	Dichter, M. A. Rat cortical neurons in cell culture: culture methods, cell morphology, electrophysiology, and synapse formation. Brain Res 149, 279-293, (1978). McCarthy, K. D. & de Vellis, J. Preparation of separate astroglial and oligodendroglial cell cultures from rat cerebral tissue. J Cell Biol 85, 890-902 (1980). Goldberg, M. P. & Choi, D. W. Combined oxygen and glucose deprivation in cortical cell culture: calcium-dependent and calcium-independent mechanisms of neuronal injury. J Neurosci 13, 3510-3524 (1993).
Ex Vivo		Organotypic Brain slice culture	Slices may be obtained from many regions of the brain, and the cultured neuronal tissue tends to retain its in vivo cytoarchatecture. This method is used to assay neuronal migration as well as other experiments requiring long term survival of neural tissue.	Gahwiler, B. H. Organotypic monolayer cultures of nervous tissue. J Neurosci Methods 4, 329-342 (1981). Gahwiler, B. H., Capogna, M., Debanne, D., McKinney, R. A. & Thompson, S. M. Organotypic slice cultures: a technique has come of age. Trends Neurosci 20, 471-477 (1997).
		Acute Brain slice culture	Used to asses the short term (minutes to hours) tissue level response to various conditions / ischemic pathologies.	Whittingham, T. S., Lust, W. D. & Passonneau, J. V. An in vitro model of ischemia: metabolic and electrical alterations in the hippocampal slice. J Neurosci 4, 793-802 (1984).

Category		Model	Notes	Duration of Ischemia	References
In Vivo	Global Ischemia	4 vessel occlusion	The vertebral arteries are permanently occluded via cauterization, and the common carotid arteries are reversibly ligated with atraumatic clamps. This model provides a reversible method of ischemia	Permanent	Eklof, B. & Siesjo, B. K. The effect of bilateral carotid artery ligation upon the blood flow and the energy state of the rat brain. Acta Physiol Scand 86, 155-165 (1972).
		2 Vessel Occlusion + hypotension	Bilaterally occluding the common carotid arteries (CCA) in association with systemic hypotension produces reversible forebrain ischemia. Blood pressure is reduced via controlled arterial hemorrhage with or without agents such as trimethaphan or phentolamine.	Transient	Pulsinelli, W. A. & Brierley, J. B. A new model of bilateral hemispheric ischemia in the unanesthetized rat. Stroke 10, 267-272 (1979).
	Focal Ischemia	Tamura Method	This method of permanent focal ischemia -involves occluding the stem of the MCA through a subtemporal craniectomy. This produces focal ischemia as well as damage localized to the frontal cortex and lateral neostriatum.	Permanent	Tamura, A., Graham, D. I., McCulloch, J. & Teasdale, G. M. Focal cerebral ischaemia in the rat: 1. Description of technique and early neuropathological consequences following middle cerebral artery occlusion. J Cereb Blood Flow Metab 1, 53-60, (1981).
		Longa Method	Longa's method involves advancing a 4-0 nylon intraluminal suture into the cervical internal carotid artery (ICA) and advancing it to occlude the MCA. It can induce ischemia without a craniectomy.	Permanent or Transient	Longa, E. Z., Weinstein, P. R., Carlson, S. & Cummins, R. Reversible middle cerebral artery occlusion without craniectomy in rats. Stroke 20, 84-91 (1989).
		Whisker Barrel Cortex Method	Three to six branches of the MCA and identified collaterals are ligated via suture to produce ischemia predominantly localized to the rat barrel cortex. Stimulation of the contralateral whiskers increases blood flow to barrel cortex, and its vasculature are visualized through intrinsic optical imaging.	Transient	Wei, L., Rovainen, C. M. & Woolsey, T. A. Ministrokes in rat barrel cortex. Stroke 26, 1459-1462 (1995).
		Photochemical / Photothrombotic Method	A photosensitive dye is injected into the brain and passes through the blood brain barrier. The dye is activated via laser and induces ischemic injury by generating microthromboses. The laser can be directed towards any area of the skull or an area of an artery wall to produce this effect.	Permanent	Eichenbaum, J. W. et al. A murine photochemical stroke model with histologic correlates of apoptotic and nonapoptotic mechanisms. J Pharmacol Toxicol Methods 47, 67-71, (2002). De Ryck, M. et al. Lubeluzole protects sensorimotor function and reduces infarct size in a photochemical stroke model in rats. J Pharmacol Exp Ther 279, 748-758 (1996).
		Endothelin-1 Method	Endothelin-1 (ET-1) is a potent vasoconstricting peptide capable of reducing cerebral blood flow and inducing transient to semi-permanent focal ischemia. when injected intracerebrally, injected adjacent to the middle cerebral artery, or topically applied to the brain. The reperfusion aspect of this technique requires characterization.	Semi-Permanent	Windle, V. et al. An analysis of four different methods of producing focal cerebral ischemia with endothelin-1 in the rat. Exp Neurol 201, 324-334, (2006).
		Autologous Clot Method	In this model a thrombus is artificially created and injected into the external carotid artery. The thromboembolism becomes lodged at the origin of the MCA and induces ischemic injury however, the location of induced ischemia as well as reperfusion are difficult to control.	Permanent	Wang, C. X., Yang, T. & Shuaib, A. An improved version of embolic model of brain ischemic injury in the rat. J Neurosci Methods 109, 147-151, (2001). Zhang, R. L., Chopp, M., Zhang, Z. G., Jiang, Q. & Ewing, J. R. A rat model of focal embolic cerebral ischemia. Brain Res 766, 83-92, (1997).

Many stroke models and experimental therapies have been developed and optimized to model different clinically situations (Table 1) [134]. The ischemic lesion is most commonly generated in these models by MCA occlusion, however the location and magnitude of the lesion can be varied by occluding one or more cerebral arteries at various locations.

B. Tracking Neuronal Progenitor Proliferation and Migration

Techniques for evaluating cell proliferation and migration in the ischemic brain have advanced significantly during the last decade. These techniques can be used in combination to analyze the infarct location, the extent of post-injury progenitor proliferation, and the direction and extent of progenitor migration.

a. 5-Bromodeoxyuridine (BrdU) Labeling

One of the most common ways to track progenitor migration is using BrdU. BrdU is a uridine analog that can be systemically administered and will incorporate into the DNA of cells undergoing the S-phase of mitosis [135, 136]. Cells labeled with BrdU can later be identified by immunofluorescence to determine proliferation at a given time period as well as patterns of migration to the infarct [127]. BrdU labeling of the DNA of mitotic cells can be passed down to the daughter cells of those originally labeled [135]. BrdU has been used in numerous studies and several methods of determining the differential extent of BrdU incorporation during DNA replication and repair have been developed [121, 137, 138]. Colocalization of BrdU with immature neuronal cell markers such as nestin or doublecortin (DCX) indicates proliferating neural precursor/migratory cells. BrdU colocalization with mature neuronal markers such as NeuN indicates that the cell had differentiated into a mature neuron. BrdU can also colocalize with endothelial or glial marker like Collagen IV and glial fibrillary acid protein (GFAP), which indicates angiogenesis and gliosis, respectively.

Like any technique, there are drawbacks to BrdU labeling. Disadvantages of using BrdU include its intrinsic property to label cells both undergoing DNA repair as well as replication, thus BrdU could label injured or dying cells [137]. Therefore, even though BrdU-labeling is a common technique, researchers may have false-positive cell counts such as BrdU-positive cells undergoing DNA damage of apoptotic death. To account for this, one would combine BrdU-labeling with other techniques like staining for cell death (TUNEL staining, for example).

BrdU may be used as a complementary label in conjunction with another label such as DCX-labeled cells or Nestin-labeled cells. These are feasible alternatives and supplements to BrdU labeling. Selective identification of proliferating cells can also be achieved by infecting a desired region such as the SVZ with a virus that labels progenitors with fluorescent tag like GFP in conjunction with the injection of BrdU [139, 140]. Cells co-expressing the viral-induced fluorescent tag and BrdU would be much more likely to have undergone mitosis during BrdU exposure as opposed to DNA repair.

b. Immunohistochemistry

The differential expression of various proteins can be visualized for the process of neurogenesis and cellular migration via immunolabeling (Fig. 3). Proteins commonly used to visualize neural progenitor migration are nestin and DCX.

Nestin is a type VI intermediate filament protein. It is an immature neuroepithelial-precursor protein expressed by migrating neural progenitors in the RMS and by progenitors migrating from the SVZ to the site of ischemic injury [5, 135, 141]. Staining for nestin expression at different time points following injury allows the analysis of the time course and magnitude of post-injury neural progenitor migration [135].

DCX is a microtubule-associated protein involved with progenitor migration [121, 142]. Progenitors in the SVZ express DCX, and it is highly expressed by migrating progenitors in the RMS in the absence of injury and is expressed in the SVZ and RMS progenitors prior to injury [135, 142, 143]. Immunolabeling following focal ischemic injury illustrates multiple chains of DCX-positive cells migrating from the SVZ to the ischemic penumbra [108, 135].

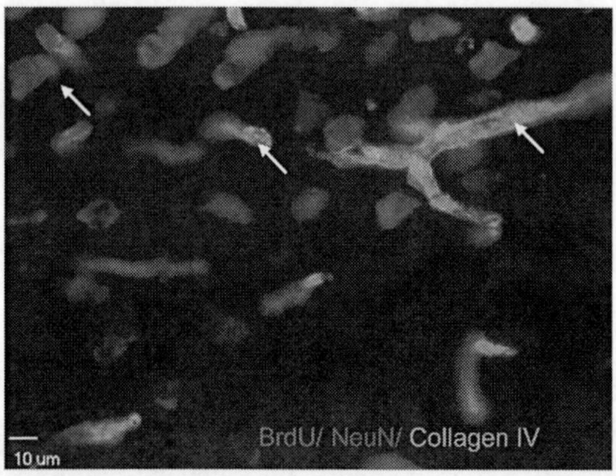

Figure 3. This is an example of immunohistochemistry in mouse brain cortical tissue: neuronal marker (NeuN, blue) and vessel marker (Collagen IV, green) colabeled with BrdU (red).

Transgenic DCX-GFP mice have been employed to examine the distribution of DCX following ischemia to indicate the extent and the source of neural progenitor proliferation108. Compared to DCX-GFP mice that did not undergo ischemia, the DCX-GFP mice with stroke revealed significantly enhanced proliferation of SVZ DCX-positive cells and the migration of these cells from the SVZ to ischemic penumbra [108, 142]. Immunofluorescence can also be used to track intrinsic cellular markers of proliferation such as proliferating cell nuclear antigen (PCNA), DCX, and nestin. These markers can also be colabeled with a GFP tag from viral vectors. However, the presence of any of these proteins can only indicate which cells were proliferating immediately prior to histological analysis, whereas BrdU can indicate which cells were proliferating at a specific time point (time of BrdU injection) [121, 136, 137].

c. Viral Vectors

Viral vectors expressing fluorescent proteins are used in tracking cellular migration. Genetically manipulated mice have been created to label the progenitor markers such as in the

Nestin-YFP mice [144, 145]. Different viral vectors can be designed based on the nature of the experiment as well as target cell type [127]. The process of using viral vectors to analyze patterns of tissue migration involves localized injection of a virus into a desired area such as the SVZ or the lateral ventricle prior to inducing ischemic injury. Because infected cells and their derivative cells will express viral DNA and hence express the inserted fluorescent protein, cellular migration from the initial area of virus injection can be tracked. Moreover, the viral genome can be modified to prevent replication *in vivo* and thus promote the localization of viral insertion [146]. The potential drawbacks of virus injections include damage mediated by the virus itself [127, 147, 148], however, the technique is continuously improving and widely recognized as accurate and reliable if viral influence is adequately controlled for [149].

X. Closing Remarks

Ischemic stroke initiates several cascades of cell death and subsequent mechanisms of trophic support and regenerative signaling for repair. Much of the endogenous repair includes rebuilding tissue, circuitry and the vascular system around the ischemic damage. This process requires the migration of endogenous neural progenitors from the SVZ. The mechanisms of this limited self-repair has been much elucidated, but nonetheless, this process itself is not sufficient for the full repair of ischemic stroke. However, understanding these endogenous mechanisms may help us enhance endogenous repair for increased repair.

References

[1] Lloyd-Jones, D., R.J. Adams, T.M. Brown, M. Carnethon, et al., Heart disease and stroke statistics--2010 update: a report from the American Heart Association. Circulation, 2010. 121(7): p. e46-e215.

[2] Lee, J.M., M.C. Grabb, G.J. Zipfel, and D.W. Choi, Brain tissue responses to ischemia. J Clin Invest, 2000. 106(6): p. 723-31.

[3] Choi, D.W., Excitotoxic cell death. J Neurobiol, 1992. 23(9): p. 1261-76.

[4] Imitola, J., K. Raddassi, K.I. Park, F.J. Mueller, et al., Directed migration of neural stem cells to sites of CNS injury by the stromal cell-derived factor 1alpha/CXC chemokine receptor 4 pathway. Proc Natl Acad Sci U S A, 2004. 101(52): p. 18117-22.

[5] Arvidsson, A., T. Collin, D. Kirik, Z. Kokaia, et al., Neuronal replacement from endogenous precursors in the adult brain after stroke. Nat Med, 2002. 8(9): p. 963-70.

[6] Li, W.L., S.P. Yu, M.E. Ogle, X.S. Ding, et al., Enhanced neurogenesis and cell migration following focal ischemia and peripheral stimulation in mice. Dev Neurobiol, 2008. 68(13): p. 1474-86.

[7] Robin, A.M., Z.G. Zhang, L. Wang, R.L. Zhang, et al., Stromal cell-derived factor 1alpha mediates neural progenitor cell motility after focal cerebral ischemia. J Cereb Blood Flow Metab, 2006. 26(1): p. 125-34.

[8] Ohab, J.J., S. Fleming, A. Blesch, and S.T. Carmichael, A neurovascular niche for neurogenesis after stroke. J Neurosci, 2006. 26(50): p. 13007-16.

[9] Lo, E.H., A new penumbra: transitioning from injury into repair after stroke. Nat Med, 2008. 14(5): p. 497-500.

[10] Choi, D.W., Glutamate neurotoxicity and diseases of the nervous system. Neuron, 1988. 1(8): p. 623-34.

[11] Lo, E.H., Experimental models, neurovascular mechanisms and translational issues in stroke research. Br J Pharmacol, 2008. 153 Suppl 1: p. S396-405.

[12] Bonfoco, E., D. Krainc, M. Ankarcrona, P. Nicotera, et al., Apoptosis and necrosis: two distinct events induced, respectively, by mild and intense insults with N-methyl-D-aspartate or nitric oxide/superoxide in cortical cell cultures. Proc Natl Acad Sci U S A, 1995. 92(16): p. 7162-6.

[13] Chu, C.T., Eaten alive: Autophagy and neuronal cell death after hypoxia-ischemia (vol 172, pg 284, 2008). American Journal of Pathology, 2008. 172(4): p. 1153-1153.

[14] Martin, L.J., N.A. Al-Abdulla, A.M. Brambrink, J.R. Kirsch, et al., Neurodegeneration in excitotoxicity, global cerebral ischemia, and target deprivation: A perspective on the contributions of apoptosis and necrosis. Brain Res Bull, 1998. 46(4): p. 281-309.

[15] Smart, I., Subependymal Layer of Mouse Brain and Its Cell Production as Shown by Radioautography after Thymidine-H3 Injection. Journal of Comparative Neurology, 1961. 116(3): p. 325-&.

[16] Altman, J., Are new neurons formed in the brains of adult mammals? Science, 1962. 135: p. 1127-8.

[17] Eriksson, P.S., E. Perfilieva, T. Bjork-Eriksson, A.M. Alborn, et al., Neurogenesis in the adult human hippocampus. Nat Med, 1998. 4(11): p. 1313-7.

[18] Gould, E., A.J. Reeves, M. Fallah, P. Tanapat, et al., Hippocampal neurogenesis in adult Old World primates. Proc Natl Acad Sci U S A, 1999. 96(9): p. 5263-7.

[19] Parent, J.M., Injury-induced neurogenesis in the adult mammalian brain. Neuroscientist, 2003. 9(4): p. 261-72.

[20] van Praag, H., B.R. Christie, T.J. Sejnowski, and F.H. Gage, Running enhances neurogenesis, learning, and long-term potentiation in mice. Proc Natl Acad Sci U S A, 1999. 96(23): p. 13427-31.

[21] van Praag, H., G. Kempermann, and F.H. Gage, Running increases cell proliferation and neurogenesis in the adult mouse dentate gyrus. Nat Neurosci, 1999. 2(3): p. 266-70.

[22] Altman, J., Autoradiographic and histological studies of postnatal neurogenesis. IV. Cell proliferation and migration in the anterior forebrain, with special reference to persisting neurogenesis in the olfactory bulb. J Comp Neurol, 1969. 137(4): p. 433-57.

[23] Petreanu, L. and A. Alvarez-Buylla, Maturation and death of adult-born olfactory bulb granule neurons: role of olfaction. J Neurosci, 2002. 22(14): p. 6106-13.

[24] Kornack, D.R. and P. Rakic, The generation, migration, and differentiation of olfactory neurons in the adult primate brain. Proc Natl Acad Sci U S A, 2001. 98(8): p. 4752-7.

[25] Chen, J., C. Zhang, H. Jiang, Y. Li, et al., Atorvastatin induction of VEGF and BDNF promotes brain plasticity after stroke in mice. J Cereb Blood Flow Metab, 2005. 25(2): p. 281-90.

[26] Belmadani, A., P.B. Tran, D. Ren, and R.J. Miller, Chemokines regulate the migration of neural progenitors to sites of neuroinflammation. J Neurosci, 2006. 26(12): p. 3182-91.

[27] Katakowski, M., J. Chen, Z.G. Zhang, M. Santra, et al., Stroke-induced subventricular zone proliferation is promoted by tumor necrosis factor-alpha-converting enzyme protease activity. J Cereb Blood Flow Metab, 2007. 27(4): p. 669-78.

[28] Jin, R., G. Yang, and G. Li, Inflammatory mechanisms in ischemic stroke: role of inflammatory cells. J Leukoc Biol, 2010. 87(5): p. 779-89.

[29] Lee, S.R., H.Y. Kim, J. Rogowska, B.Q. Zhao, et al., Involvement of matrix metalloproteinase in neuroblast cell migration from the subventricular zone after stroke. J Neurosci, 2006. 26(13): p. 3491-5.

[30] Cybulska-Klosowicz, A., M. Liguz-Lecznar, D. Nowicka, M. Ziemka-Nalecz, et al., Matrix metalloproteinase inhibition counteracts impairment of cortical experience-dependent plasticity after photothrombotic stroke. Eur J Neurosci, 2011. 33(12): p. 2238-46.

[31] Kang, S.S., J.H. Kook, S. Hwang, S.H. Park, et al., Inhibition of matrix metalloproteinase-9 attenuated neural progenitor cell migration after photothrombotic ischemia. Brain Res, 2008. 1228: p. 20-6.

[32] Widera, D., W. Holtkamp, F. Entschladen, B. Niggemann, et al., MCP-1 induces migration of adult neural stem cells. Eur J Cell Biol, 2004. 83(8): p. 381-7.

[33] Yan, Y.P., K.A. Sailor, B.T. Lang, S.W. Park, et al., Monocyte chemoattractant protein-1 plays a critical role in neuroblast migration after focal cerebral ischemia. J Cereb Blood Flow Metab, 2007. 27(6): p. 1213-24.

[34] Zhang, R.L., Z.G. Zhang, L. Zhang, and M. Chopp, Proliferation and differentiation of progenitor cells in the cortex and the subventricular zone in the adult rat after focal cerebral ischemia. Neuroscience, 2001. 105(1): p. 33-41.

[35] Zhang, R., Z. Zhang, C. Zhang, L. Zhang, et al., Stroke transiently increases subventricular zone cell division from asymmetric to symmetric and increases neuronal differentiation in the adult rat. J Neurosci, 2004. 24(25): p. 5810-5.

[36] Kuge, A., S. Takemura, Y. Kokubo, S. Sato, et al., Temporal profile of neurogenesis in the subventricular zone, dentate gyrus and cerebral cortex following transient focal cerebral ischemia. Neurol Res, 2009. 31(9): p. 969-76.

[37] Arvidsson, A., Z. Kokaia, and O. Lindvall, N-methyl-D-aspartate receptor-mediated increase of neurogenesis in adult rat dentate gyrus following stroke. Eur J Neurosci, 2001. 14(1): p. 10-8.

[38] Zhang, Z., K. Davies, J. Prostak, J. Fenstermacher, et al., Quantitation of microvascular plasma perfusion and neuronal microtubule-associated protein in ischemic mouse brain by laser-scanning confocal microscopy. J Cereb Blood Flow Metab,1999. 19(1): p. 68-78.

[39] Wang, L., M. Chopp, S.R. Gregg, R.L. Zhang, et al., Neural progenitor cells treated with EPO induce angiogenesis through the production of VEGF. J Cereb Blood Flow Metab, 2008. 28(7): p. 1361-8.

[40] Zhou, Y., P.H. Larsen, C. Hao, and V.W. Yong, CXCR4 is a major chemokine receptor on glioma cells and mediates their survival. J Biol Chem, 2002. 277(51): p. 49481-7.

[41] Rubin, J.B., A.L. Kung, R.S. Klein, J.A. Chan, et al., A small-molecule antagonist of CXCR4 inhibits intracranial growth of primary brain tumors. Proc Natl Acad Sci U S A, 2003. 100(23): p. 13513-8.

[42] Sun, L., J. Lee, and H.A. Fine, Neuronally expressed stem cell factor induces neural stem cell migration to areas of brain injury. J Clin Invest, 2004. 113(9): p. 1364-74.

[43] Thored, P., A. Arvidsson, E. Cacci, H. Ahlenius, et al., Persistent production of neurons from adult brain stem cells during recovery after stroke. Stem Cells, 2006. 24(3): p. 739-47.

[44] Ji, J.F., B.P. He, S.T. Dheen, and S.S. Tay, Expression of chemokine receptors CXCR4, CCR2, CCR5 and CX3CR1 in neural progenitor cells isolated from the subventricular zone of the adult rat brain. Neurosci Lett, 2004. 355(3): p. 236-40.

[45] Ekdahl, C.T., Z. Kokaia, and O. Lindvall, Brain inflammation and adult neurogenesis: the dual role of microglia. Neuroscience, 2009. 158(3): p. 1021-9.

[46] Covey, M.V., D. Loporchio, K.D. Buono, and S.W. Levison, Opposite effect of inflammation on subventricular zone versus hippocampal precursors in brain injury. Ann Neurol, 2011.

[47] Aarum, J., K. Sandberg, S.L. Haeberlein, and M.A. Persson, Migration and differentiation of neural precursor cells can be directed by microglia. Proc Natl Acad Sci U S A, 2003. 100(26): p. 15983-8.

[48] Gebicke-Haerter, P.J., O. Spleiss, L.Q. Ren, H. Li, et al., Microglial chemokines and chemokine receptors. Prog Brain Res, 2001. 132: p. 525-32.

[49] Streit, W.J., Microglia as neuroprotective, immunocompetent cells of the CNS. Glia, 2002. 40(2): p. 133-9.

[50] Li, M. and R.M. Ransohoff, Multiple roles of chemokine CXCL12 in the central nervous system: a migration from immunology to neurobiology. Prog Neurobiol, 2008. 84(2): p. 116-31.

[51] Matthys, P., S. Hatse, K. Vermeire, A. Wuyts, et al., AMD3100, a potent and specific antagonist of the stromal cell-derived factor-1 chemokine receptor CXCR4, inhibits autoimmune joint inflammation in IFN-gamma receptor-deficient mice. J Immunol, 2001. 167(8): p. 4686-92.

[52] Bleul, C.C., R.C. Fuhlbrigge, J.M. Casasnovas, A. Aiuti, et al., A highly efficacious lymphocyte chemoattractant, stromal cell-derived factor 1 (SDF-1). J Exp Med, 1996. 184(3): p. 1101-9.

[53] Ma, Q., D. Jones, P.R. Borghesani, R.A. Segal, et al., Impaired B-lymphopoiesis, myelopoiesis, and derailed cerebellar neuron migration in CXCR4- and SDF-1-deficient mice. Proc Natl Acad Sci U S A, 1998. 95(16): p. 9448-53.

[54] Cruz-Orengo, L., D.W. Holman, D. Dorsey, L. Zhou, et al., CXCR7 influences leukocyte entry into the CNS parenchyma by controlling abluminal CXCL12 abundance during autoimmunity. Journal of Experimental Medicine, 2011. 208(2): p. 327-339.

[55] Huising, M.O., R.J. Stet, C.P. Kruiswijk, H.F. Savelkoul, et al., Molecular evolution of CXC chemokines: extant CXC chemokines originate from the CNS. Trends Immunol, 2003. 24(6): p. 307-13.

[56] Zou, Y.R., A.H. Kottmann, M. Kuroda, I. Taniuchi, et al., Function of the chemokine receptor CXCR4 in haematopoiesis and in cerebellar development. Nature, 1998. 393(6685): p. 595-9.

[57] Rosu-Myles, M., L. Gallacher, B. Murdoch, D.A. Hess, et al., The human hematopoietic stem cell compartment is heterogeneous for CXCR4 expression. Proc Natl Acad Sci U S A, 2000. 97(26): p. 14626-31.

[58] Pituch-Noworolska, A., M. Majka, A. Janowska-Wieczorek, M. Baj-Krzyworzeka, et al., Circulating CXCR4-positive stem/progenitor cells compete for SDF-1-positive niches in bone marrow, muscle and neural tissues: an alternative hypothesis to stem cell plasticity. Folia Histochem Cytobiol, 2003. 41(1): p. 13-21.

[59] Hatch, H.M., D. Zheng, M.L. Jorgensen, and B.E. Petersen, SDF-1alpha/CXCR4: a mechanism for hepatic oval cell activation and bone marrow stem cell recruitment to the injured liver of rats. Cloning Stem Cells, 2002. 4(4): p. 339-51.

[60] Crane, I.J., C.A. Wallace, S. McKillop-Smith, and J.V. Forrester, CXCR4 receptor expression on human retinal pigment epithelial cells from the blood-retina barrier leads to chemokine secretion and migration in response to stromal cell-derived factor 1 alpha. J Immunol, 2000. 165(8): p. 4372-8.

[61] Guo, Y., G. Hangoc, H. Bian, L.M. Pelus, et al., SDF-1/CXCL12 enhances survival and chemotaxis of murine embryonic stem cells and production of primitive and definitive hematopoietic progenitor cells. Stem Cells, 2005. 23(9): p. 1324-32.

[62] Zhu, Y., T. Yu, X.C. Zhang, T. Nagasawa, et al., Role of the chemokine SDF-1 as the meningeal attractant for embryonic cerebellar neurons. Nat Neurosci, 2002. 5(8): p. 719-20.

[63] Liu, X.S., M. Chopp, M. Santra, A. Hozeska-Solgot, et al., Functional response to SDF1 alpha through over-expression of CXCR4 on adult subventricular zone progenitor cells. Brain Res, 2008. 1226: p. 18-26.

[64] Burns, J.M., B.C. Summers, Y. Wang, A. Melikian, et al., A novel chemokine receptor for SDF-1 and I-TAC involved in cell survival, cell adhesion, and tumor development. Journal of Experimental Medicine, 2006. 203(9): p. 2201-2213.

[65] Li, W.L., S.P. Yu, M.E. Ogle, X.S. Ding, et al., Enhanced Neurogenesis and Cell Migration Following Focal Ischemia and Peripheral Stimulation in Mice. Developmental Neurobiology, 2008. 68(13): p. 1474-1486.

[66] Hasselblatt, M., A. Jeibmann, B. Riesmeier, D. Maintz, et al., Granulocyte-colony stimulating factor (G-CSF) and G-CSF receptor expression in human ischemic stroke. Acta Neuropathologica, 2007. 113(1): p. 45-51.

[67] Stumm, R.K., J. Rummel, V. Junker, C. Culmsee, et al., A dual role for the SDF-1/CXCR4 chemokine receptor system in adult brain: Isoform-selective regulation of SDF-1 expression modulates CXCR4-dependent neuronal plasticity and cerebral leukocyte recruitment after focal ischemia. Journal of Neuroscience, 2002. 22(14): p. 5865-5878.

[68] Miller, J.T., J.H. Bartley, H.J. Wimborne, A.L. Walker, et al., The neuroblast and angioblast chemotaxic factor SDF-1 (CXCL12) expression is briefly up regulated by reactive astrocytes in brain following neonatal hypoxic-ischemic injury. BMC Neurosci, 2005. 6: p. 63.

[69] Schonemeier, B., S. Schulz, V. Hoellt, and R. Stumm, Enhanced expression of the CXCl12/SDF-1 chemokine receptor CXCR7 after cerebral ischemia in the rat brain. J Neuroimmunol, 2008. 198(1-2): p. 39-45.

[70] Shyu, W.C., S.Z. Lin, P.S. Yen, C.Y. Su, et al., Stromal cell-derived factor-1 alpha promotes neuroprotection, angiogenesis, and mobilization/homing of bone marrow-derived cells in stroke rats. J Pharmacol Exp Ther, 2008. 324(2): p. 834-49.

[71] Teicher, B.A. and S.P. Fricker, CXCL12 (SDF-1)/CXCR4 pathway in cancer. Clin Cancer Res, 2010. 16(11): p. 2927-31.

[72] 72. Vlahakis, S.R., A. Villasis-Keever, T. Gomez, M. Vanegas, et al., G protein-coupled chemokine receptors induce both survival and apoptotic signaling pathways. J Immunol, 2002. 169(10): p. 5546-54.

[73] Ceradini, D.J., A.R. Kulkarni, M.J. Callaghan, O.M. Tepper, et al., Progenitor cell trafficking is regulated by hypoxic gradients through HIF-1 induction of SDF-1. Nature Medicine, 2004. 10(8): p. 858-864.

[74] Petit, I., M. Szyper-Kravitz, A. Nagler, M. Lahav, et al., G-CSF induces stem cell mobilization by decreasing bone marrow SDF-1 and up-regulating CXCR4. Nat Immunol, 2002. 3(7): p. 687-94.

[75] Ferrara, N., H.P. Gerber, and J. LeCouter, The biology of VEGF and its receptors. Nat Med, 2003. 9(6): p. 669-76.

[76] Tammela, T., B. Enholm, K. Alitalo, and K. Paavonen, The biology of vascular endothelial growth factors. Cardiovasc Res, 2005. 65(3): p. 550-63.

[77] Keck, P.J., S.D. Hauser, G. Krivi, K. Sanzo, et al., Vascular permeability factor, an endothelial cell mitogen related to PDGF. Science, 1989. 246(4935): p. 1309-12.

[78] Leung, D.W., G. Cachianes, W.J. Kuang, D.V. Goeddel, et al., Vascular endothelial growth factor is a secreted angiogenic mitogen. Science, 1989. 246(4935): p. 1306-9.

[79] Neufeld, G., T. Cohen, S. Gengrinovitch, and Z. Poltorak, Vascular endothelial growth factor (VEGF) and its receptors. Faseb Journal, 1999. 13(1): p. 9-22.

[80] Sun, Y., K. Jin, L. Xie, J. Childs, et al., VEGF-induced neuroprotection, neurogenesis, and angiogenesis after focal cerebral ischemia. J Clin Invest, 2003. 111(12): p. 1843-51.

[81] Wang, Y., K. Jin, X.O. Mao, L. Xie, et al., VEGF-overexpressing transgenic mice show enhanced post-ischemic neurogenesis and neuromigration. J Neurosci Res, 2007. 85(4): p. 740-7.

[82] Jin, K.L., X.O. Mao, and D.A. Greenberg, Vascular endothelial growth factor: direct neuroprotective effect in in vitro ischemia. Proc Natl Acad Sci U S A, 2000. 97(18): p. 10242-7.

[83] Whitaker, V.R., L. Cui, S. Miller, S.P. Yu, et al., Whisker stimulation enhances angiogenesis in the barrel cortex following focal ischemia in mice. Journal of Cerebral Blood Flow and Metabolism, 2007. 27(1): p. 57-68.

[84] Wei, L., J.P. Erinjeri, C.M. Rovainen, and T.A. Woolsey, Collateral growth and angiogenesis around cortical stroke. Stroke, 2001. 32(9): p. 2179-84.

[85] Li, W.L., J.L. Fraser, S.P. Yu, J. Zhu, et al., The role of VEGF/VEGF2 signaling in peripheral stimulation-induced cerebral neurovascular regeneration after ischemic stroke in mice. Exp Brain Res, 2011. in press.

[86] Zhang, H., L. Vutskits, M.S. Pepper, and J.Z. Kiss, VEGF is a chemoattractant for FGF-2-stimulated neural progenitors. J Cell Biol, 2003. 163(6): p. 1375-84.

[87] Olsson, A.K., A. Dimberg, J. Kreuger, and L. Claesson-Welsh, VEGF receptor signalling - in control of vascular function. Nat Rev Mol Cell Biol, 2006. 7(5): p. 359-71.

[88] Jin, K., Y. Zhu, Y. Sun, X.O. Mao, et al., Vascular endothelial growth factor (VEGF) stimulates neurogenesis in vitro and in vivo. Proc Natl Acad Sci U S A, 2002. 99(18): p. 11946-50.

[89] Zhang, Z.G., L. Zhang, W. Tsang, H. Soltanian-Zadeh, et al., Correlation of VEGF and angiopoietin expression with disruption of blood-brain barrier and angiogenesis after focal cerebral ischemia. J Cereb Blood Flow Metab, 2002. 22(4): p. 379-92.

[90] Wang, Y.Q., X. Guo, M.H. Qiu, X.Y. Feng, et al., VEGF overexpression enhances striatal neurogenesis in brain of adult rat after a transient middle cerebral artery occlusion. J Neurosci Res, 2007. 85(1): p. 73-82.

[91] Zhang, Z.G., L. Zhang, Q. Jiang, R. Zhang, et al., VEGF enhances angiogenesis and promotes blood-brain barrier leakage in the ischemic brain. J Clin Invest, 2000. (7): p. 829-38.

[92] Bath, K.G., N. Mandairon, D. Jing, R. Rajagopal, et al., Variant brain-derived neurotrophic factor (Val66Met) alters adult olfactory bulb neurogenesis and spontaneous olfactory discrimination. J Neurosci, 2008. 28(10): p. 2383-93.

[93] Zigova, T., V. Pencea, S.J. Wiegand, and M.B. Luskin, Intraventricular administration of BDNF increases the number of newly generated neurons in the adult olfactory bulb. Molecular and cellular neurosciences, 1998. 11(4): p. 234-45.

[94] Pencea, V., K.D. Bingaman, S.J. Wiegand, and M.B. Luskin, Infusion of brain-derived neurotrophic factor into the lateral ventricle of the adult rat leads to new neurons in the parenchyma of the striatum, septum, thalamus, and hypothalamus. The Journal of neuroscience : the official journal of the Society for Neuroscience, 2001. 21(17): p. 6706-17.

[95] Galvao, R.P., J.M. Garcia-Verdugo, and A. Alvarez-Buylla, Brain-derived neurotrophic factor signaling does not stimulate subventricular zone neurogenesis in adult mice and rats. J Neurosci, 2008. 28(50): p. 13368-83.

[96] Schabitz, W.R., C. Sommer, W. Zoder, M. Kiessling, et al., Intravenous brain-derived neurotrophic factor reduces infarct size and counterregulates Bax and Bcl-2 expression after temporary focal cerebral ischemia. Stroke; a journal of cerebral circulation, 2000. 31(9): p. 2212-7.

[97] Kirschenbaum, B. and S.A. Goldman, Brain-derived neurotrophic factor promotes the survival of neurons arising from the adult rat forebrain subependymal zone. Proc Natl Acad Sci U S A, 1995. 92(1): p. 210-4.

[98] Schabitz, W.R., S. Schwab, M. Spranger, and W. Hacke, Intraventricular brain-derived neurotrophic factor reduces infarct size after focal cerebral ischemia in rats. J Cereb Blood Flow Metab, 1997. 17(5): p. 500-6.

[99] Comelli, M.C., M.S. Seren, D. Guidolin, R.M. Manev, et al., Photochemical stroke and brain-derived neurotrophic factor (BDNF) mRNA expression. Neuroreport, 1992. 3(6): p. 473-6.

[100] Borghesani, P.R., J.M. Peyrin, R. Klein, J. Rubin, et al., BDNF stimulates migration of cerebellar granule cells. Development, 2002. 129(6): p. 1435-42.

[101] Behar, T.N., M.M. Dugich-Djordjevic, Y.X. Li, W. Ma, et al., Neurotrophins stimulate chemotaxis of embryonic cortical neurons. Eur J Neurosci, 1997. 9(12): p. 2561-70.

[102] Chiaramello, S., G. Dalmasso, L. Bezin, D. Marcel, et al., BDNF/ TrkB interaction regulates migration of SVZ precursor cells via PI3-K and MAP-K signalling pathways. The European journal of neuroscience, 2007. 26(7): p. 1780-90.

[103] Pencea, V., K.D. Bingaman, S.J. Wiegand, and M.B. Luskin, Infusion of brain-derived neurotrophic factor into the lateral ventricle of the adult rat leads to new neurons in the parenchyma of the striatum, septum, thalamus, and hypothalamus. Journal of Neuroscience, 2001. 21(17): p. 6706-6717.

[104] Benraiss, A., E. Chmielnicki, K. Lerner, D. Roh, et al., Adenoviral brain-derived neurotrophic factor induces both neostriatal and olfactory neuronal recruitment from endogenous progenitor cells in the adult forebrain. J Neurosci, 2001. 21(17): p. 6718-31.

[105] Jacques, T.S., J.B. Relvas, S. Nishimura, R. Pytela, et al., Neural precursor cell chain migration and division are regulated through different beta1 integrins. Development, 1998. 125(16): p. 3167-77.

[106] Rakic, P., Principles of neural cell migration. Experientia, 1990. 46(9): p. 882-91.

[107] Lois, C., J.M. Garcia-Verdugo, and A. Alvarez-Buylla, Chain migration of neuronal precursors. Science, 1996. 271(5251): p. 978-81.

[108] Zhang, R.L., M. Chopp, S.R. Gregg, Y. Toh, et al., Patterns and dynamics of subventricular zone neuroblast migration in the ischemic striatum of the adult mouse. J Cereb Blood Flow Metab, 2009. 29(7): p. 1240-50.

[109] Bovetti, S., Y.C. Hsieh, P. Bovolin, I. Perroteau, et al., Blood vessels form a scaffold for neuroblast migration in the adult olfactory bulb. J Neurosci, 2007. 27(22): p. 5976-80.

[110] Thored, P., J. Wood, A. Arvidsson, J. Cammenga, et al., Long-term neuroblast migration along blood vessels in an area with transient angiogenesis and increased vascularization after stroke. Stroke, 2007. 38(11): p. 3032-9.

[111] Louissaint, A., Jr., S. Rao, C. Leventhal, and S.A. Goldman, Coordinated interaction of neurogenesis and angiogenesis in the adult songbird brain. Neuron, 2002. 34(6): p. 945-60.

[112] Abumiya, T., J. Lucero, J.H. Heo, M. Tagaya, et al., Activated microvessels express vascular endothelial growth factor and integrin alpha(v)beta3 during focal cerebral ischemia. J Cereb Blood Flow Metab, 1999. 19(9): p. 1038-50.

[113] Cox, S.B., T.A. Woolsey, and C.M. Rovainen, Localized dynamic changes in cortical blood flow with whisker stimulation corresponds to matched vascular and neuronal architecture of rat barrels. J Cereb Blood Flow Metab, 1993. 13(6): p. 899-913.

[114] Whitaker, V.R., L. Cui, S. Miller, S.P. Yu, et al., Whisker stimulation enhances angiogenesis in the barrel cortex following focal ischemia in mice. J Cereb Blood Flow Metab, 2007. 27(1): p. 57-68.

[115] Hallett, H., L. Churchill, P. Taishi, A. De, et al., Whisker stimulation increases expression of nerve growth factor- and interleukin-1beta-immunoreactivity in the rat somatosensory cortex. Brain Res, 2010. 1333: p. 48-56.

[116] Hicks, A.U., K. Hewlett, V. Windle, G. Chernenko, et al., Enriched environment enhances transplanted subventricular zone stem cell migration and functional recovery after stroke. Neuroscience, 2007. 146(1): p. 31-40.

[117] Lipton, P., Ischemic cell death in brain neurons. Physiol Rev, 1999. 79(4): p. 568.

[118] Memezawa, H., H. Minamisawa, M.L. Smith, and B.K. Siesjo, Ischemic penumbra in a model of reversible middle cerebral artery occlusion in the rat. Exp Brain Res, 1992. 89(1): p. 67-78.

[119] Taoufik, E. and L. Probert, Ischemic neuronal damage. Curr Pharm Des, 2008. 14(33): p. 3565-73.
[120] Yuan, J., Neuroprotective strategies targeting apoptotic and necrotic cell death for stroke. Apoptosis, 2009. 14(4): p. 469-77.
[121] Jin, K., X. Wang, L. Xie, X.O. Mao, et al., Evidence for stroke-induced neurogenesis in the human brain. Proc Natl Acad Sci U S A, 2006. 103(35): p. 13198-202.
[122] Gahwiler, B.H., M. Capogna, D. Debanne, R.A. McKinney, et al., Organotypic slice cultures: a technique has come of age. Trends Neurosci, 1997. 20(10): p. 471-7.
[123] 123. Maretzky, T., A. Evers, W. Zhou, S.L. Swendeman, et al., Migration of growth factor-stimulated epithelial and endothelial cells depends on EGFR transactivation by ADAM17. Nat Commun, 2011. 2: p. 229.
[124] Durbec, P., I. Franceschini, F. Lazarini, and M. Dubois-Dalcq, In vitro migration assays of neural stem cells. Methods Mol Biol, 2008. 438: p. 213-25.
[125] Wei, J.F., L. Wei, X. Zhou, Z.Y. Lu, et al., Formation of Kv2.1-FAK complex as a mechanism of FAK activation, cell polarization and enhanced motility. J Cell Physiol, 2008. 217(2): p. 544-57.
[126] Gahwiler, B.H., Organotypic monolayer cultures of nervous tissue. J Neurosci Methods, 1981. 4(4): p. 329-42.
[127] Lossi, L., S. Alasia, C. Salio, and A. Merighi, Cell death and proliferation in acute slices and organotypic cultures of mammalian CNS. Prog Neurobiol, 2009. 88(4): p. 221-45.
[128] Chopp, M., Z.G. Zhang, and Q. Jiang, Neurogenesis, angiogenesis, and MRI indices of functional recovery from stroke. Stroke, 2007. 38(2 Suppl): p. 827-31.
[129] Reith, J., H.S. Jorgensen, P.M. Pedersen, H. Nakayama, et al., Body temperature in acute stroke: relation to stroke severity, infarct size, mortality, and outcome. Lancet, 1996. 347(8999): p. 422-5.
[130] Vannucci, S.J., L.B. Willing, S. Goto, N.J. Alkayed, et al., Experimental stroke in the female diabetic, db/db, mouse. J Cereb Blood Flow Metab, 2001. 21(1): p. 52-60.
[131] Duverger, D. and E.T. MacKenzie, The quantification of cerebral infarction following focal ischemia in the rat: influence of strain, arterial pressure, blood glucose concentration, and age. J Cereb Blood Flow Metab, 1988. 8(4): p. 449-61.
[132] Kawamura, S., N. Yasui, M. Shirasawa, and H. Fukasawa, Rat middle cerebral artery occlusion using an intraluminal thread technique. Acta Neurochir (Wien), 1991. 109(3-4): p. 126-32.
[133] Graham, S.M., L.D. McCullough, and S.J. Murphy, Animal models of ischemic stroke: balancing experimental aims and animal care. Comp Med, 2004. 54(5): p. 486-96.
[134] Liu, S., G. Zhen, B.P. Meloni, K. Campbell, et al., Rodent Stroke Model Guidelines for Preclinical Stroke Trials (1st Edition). J Exp Stroke Transl Med, 2009. 2(2): p. 2-27.
[135] Jin, K., Y. Sun, L. Xie, A. Peel, et al., Directed migration of neuronal precursors into the ischemic cerebral cortex and striatum. Mol Cell Neurosci, 2003. 24(1): p. 171-89.
[136] Li, J., Z. Lu, W.L. Li, S.P. Yu, et al., Cell death and proliferation in NF-kappaB p50 knockout mouse after cerebral ischemia. Brain Res, 2008. 1230: p. 281-9.
[137] Dolbeare, F., Bromodeoxyuridine: a diagnostic tool in biology and medicine, Part III. Proliferation in normal, injured and diseased tissue, growth factors, differentiation, DNA replication sites and in situ hybridization. Histochem J, 1996. 28(8): p. 531-75.

[138] Jin, K., X.O. Mao, Y. Sun, L. Xie, et al., Stem cell factor stimulates neurogenesis in vitro and in vivo. J Clin Invest, 2002. 110(3): p. 311-9.

[139] Kay, M.A., J.C. Glorioso, and L. Naldini, Viral vectors for gene therapy: the art of turning infectious agents into vehicles of therapeutics. Nat Med, 2001. 7(1): p. 33-40.

[140] 140. Mozdziak, P. and E. Schultz, Retroviral labeling is an appropriate marker for dividing cells. Biotech Histochem, 2000. 75(3): p. 141-6.

[141] Tanvig, M., M. Blaabjerg, R.K. Andersen, A. Villa, et al., A brain slice culture model for studies of endogenous and exogenous precursor cell migration in the rostral migratory stream. Brain Res, 2009. 1295: p. 1-12.

[142] Walker, T.L., T. Yasuda, D.J. Adams, and P.F. Bartlett, The doublecortin-expressing population in the developing and adult brain contains multipotential precursors in addition to neuronal-lineage cells. J Neurosci, 2007. 27(14): p. 3734-42.

[143] Parent, J.M., V.V. Valentin, and D.H. Lowenstein, Prolonged seizures increase proliferating neuroblasts in the adult rat subventricular zone-olfactory bulb pathway. J Neurosci, 2002. 22(8): p. 3174-88.

[144] Lagace, D.C., M.C. Whitman, M.A. Noonan, J.L. Ables, et al., Dynamic contribution of nestin-expressing stem cells to adult neurogenesis. J Neurosci, 2007. 27(46): p. 12623-9.

[145] Beech, R.D., M.A. Cleary, H.B. Treloar, A.J. Eisch, et al., Nestin promoter/enhancer directs transgene expression to precursors of adult generated periglomerular neurons. J Comp Neurol, 2004. 475(1): p. 128-41.

[146] Doetsch, F., I. Caille, D.A. Lim, J.M. Garcia-Verdugo, et al., Subventricular zone astrocytes are neural stem cells in the adult mammalian brain. Cell, 1999. 97(6): p. 703-16.

[147] Falk, A., N. Holmstrom, M. Carlen, R. Cassidy, et al., Gene delivery to adult neural stem cells. Exp Cell Res, 2002. 279(1): p. 34-9.

[148] Hughes, S.M., F. Moussavi-Harami, S.L. Sauter, and B.L. Davidson, Viral-mediated gene transfer to mouse primary neural progenitor cells. Mol Ther, 2002. 5(1): p. 16-24.

[149] Ehrengruber, M.U., S. Hennou, H. Bueler, H.Y. Naim, et al., Gene transfer into neurons from hippocampal slices: Comparison of recombinant Semliki Forest virus, adenovirus, adeno-associated virus, lentivirus, and measles virus. Molecular and Cellular Neuroscience, 2001. 17(5): p. 855-871.

In: Neuronal Migration
Editors: Andrew Girard and Luca Moreau
ISBN: 978-1-62100-059-4

Chapter VI

The Influence of the Basement Membrane on Neuronal Positioning in the Cerebral Cortex

Margareta Nikolić[1] and Susan C. Brown[2]
[1]Department of Medicine, Imperial College, London, UK
[2]Department of Veterinary Basic Sciences, Royal Veterinary College, University of London, UK

Abstract

Correct neuronal migration is a key event for normal development and functioning of the central nervous system (CNS). In the mammalian cerebral cortex neurones ultimately arrange into a complex laminar configuration consisting of six distinguishable layers. Much has been discovered concerning the molecular and cellular mechanisms that underlie the correct movement of immature neurones within the cerebral cortex, as well as the crucial role played by the radial glial scaffold. In recent years, the importance of the interaction between the radial glial end-feet and the pial basement membrane has become increasingly apparent, and it is this aspect which forms the focus of the present review. We begin by outlining some of the key processes involved in the organisation of the cerebral cortex, focusing on several recent breakthroughs particularly the role of cytoskeletal and signalling proteins. We discuss basement membranes, specifically the role played by the pial basement membrane in maintaining the correct laminar structure of the developing cortex. Finally we end with a discussion on how animal models have contributed to our understanding of cortical malformations, such as cobblestone lissencephaly and speculate on developments that may ultimately offer avenues for future therapeutic interventions.

FORMATION AND STRUCTURAL ORGANISATION OF THE CEREBRAL CORTEX

In the mammalian forebrain (telencephalon), neocortical neurones assemble along the radial axis into a six layered structure that constitutes the cortical plate. Incoming and outgoing axons form the white mater-rich intermediate zone (IZ) which is evident immediately ventral to the cortical plate. Beneath the IZ lies the germinal zone of the dorsal telencephalon (Merot et al., 2009; Rakic, 2006). Most of the neurones that reside in the cerebral cortex originate from two major germinal zones (Figure 1A). Glutamatergic projection neurones arise in the germinal layers of the dorsal telencephalon and migrate radially towards the pial surface (Merot et al., 2009). Most GABAergic interneurones arise from the ventral telencephalon and migrate tangentially towards and radially within the cortex (Marin and Rubenstein, 2003; Metin et al., 2006). Despite originating from different germinal zones, interneurones and pyramidal neurones born at the same time ultimately co-reside in a cortical layer (Valcanis and Tan, 2003). This remarkable fact illustrates the degree of coordination that underpins the normal formation of the cerebral cortex.

The neocortex is formed in an inside-out manner, where neurones born early populate deep layers and those born late reside more superficially (Figure 1B) (Ayala et al., 2007; Tanaka et al., 2006). A major determinant in the correct laminar organisation of the neocortex is the ability of immature postmitotic neurones to migrate in a timely manner from the germinal zones into the cortical plate. This process depends on the ability of immature neurones to pass through previously formed layers and stop migrating at peripheral positions, close to the pial surface. It begins when the first region of postmitotic neurones, the primordial plexiform layer (preplate; PP) is split into the marginal zone (MZ) and subplate (SP) by migrating layer VI neurones. These are subsequently passed by younger glutamatergic neurones which will sequentially establish layers V, IV, III and II (Ayala et al., 2007; Merot et al., 2009). Any alterations in the rate and direction of neuronal movement can cause laminar disorganisation and a deviation from the normal six layered structure, which is frequently found associated with a spectrum of developmental problems spanning from mild learning difficulties to foetal mortality (Figure 1C) (Bielas et al., 2004; Cayre et al., 2009; Marin and Rubenstein, 2003; Metin et al., 2008; Mochida and Walsh, 2004; Nakajima, 2007; Valiente and Marin, 2010).

Radial glia (RG) are one of the most important cells in the developing neocortex. RG soma reside in the germinal zone of the dorsal telencephalon, where they undergo cell cycle-dependent interkinetic movements (Del Bene, 2011; Gotz and Huttner, 2005). Significantly, RG extend long basal processes across the cortical plate, the end-feet of which contact the pial basement membrane (BM) and glial limitans at the surface of the brain. In this manner RG form an essential scaffold, supporting the migration of immature glutamatergic neurones (Merot et al., 2009; Rakic, 2003). Consequently, alterations of the glial scaffold are known to cause disruptions in the positioning of cortical glutamatergic neurones (Rakic, 2003). Interestingly, tangential migration of interneurones into the cortical plate was until recently thought to occur independently of radial glia. However, improvements in cell specific labelling and live imaging have suggested close interactions between migrating GABAergic interneurones and the radial processes or end-feet of RG. (Yokota et al., 2007).

It is therefore likely that RG affect the positioning of cortical interneurones through direct contact as well as indirectly via the correct positioning of glutamatergic neurones.

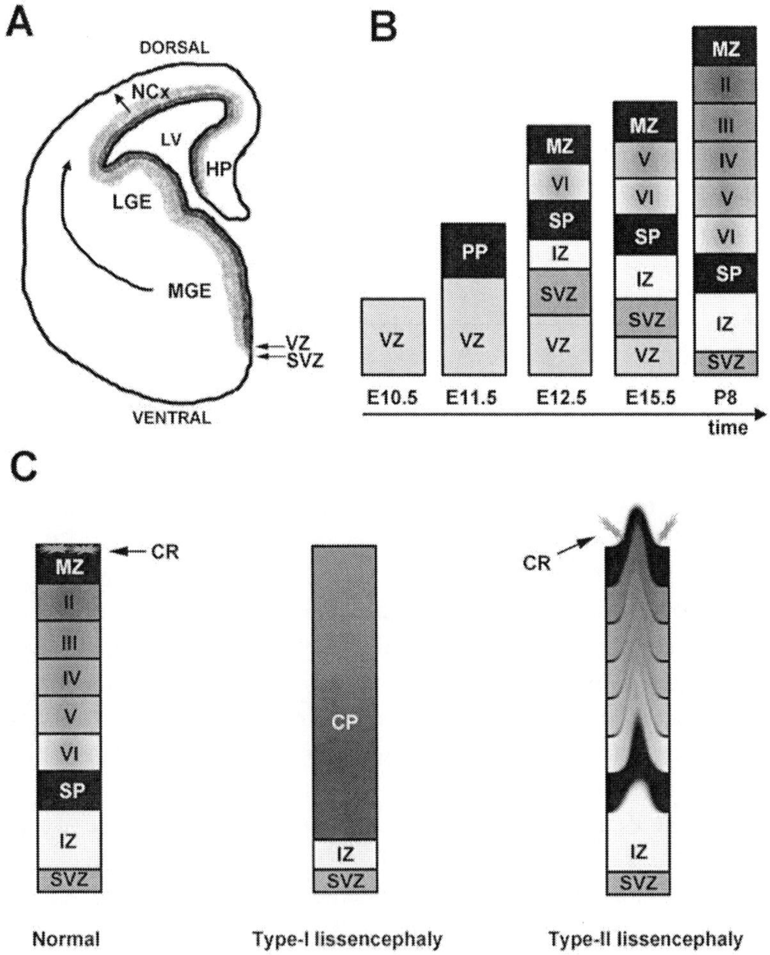

Figure 1. (A) Most neurones that populate the neocortex originate from two germinal zones. Glutamatergic pyramidal neurones migrate radially from the germinal zone in the dorsal telencephalon, while GABAergic interneurones migrate tangentially primarily from the medial ganglionic eminence (MGE) towards the neocortex (migratory routes shown with arrows). The germinal zone consists of two regions, the ventricular zone (VZ) and subventricular zone (SVZ), as shown with arrows. Abbreviations, NCx (Neocortex); LV (lateral Ventricle); MGE (Medial ganglionic eminence); LGE (Lateral ganglionic eminence); HP (Hippocampus). (B) Radial migration of glutamatergic neurones follows a defined pattern during embryogenesis. Developmental stages are given in mice where E stands for *embryonic* and P for *postnatal.* At 10.5 days of gestation (E10.5) the neural tube is lined with a proliferating germinal zone (VZ), which primarily acts to expand the proliferative pool of progenitors in addition to generating early cortical neurones. The Preplate (PP) contains the first postmitotic neurones of the cerebral cortex. It is split into a peripheral marginal zone (MZ) and deep subplate (SP) region by migrating neurones that originate from the VZ of the dorsal telencephalon and will form layer VI of the cortical plate. Subsequent layers V, IV, III and II will migrate past the SP and previously formed layers to generate a new laminar structure beneath the MZ. (C) Diagrammatic representations of early postnatal brains that are normally formed or originate from type-I and type-II lissencephalies. The position and displacement of Cajal Retzius cells (CR) are shown. Type-I lissencephaly is characterised by a reduced and disrupted white matter and disorganised cortical plate in which the classical six layered structure is not apparent. Type-II (cobblestone) lissencephaly features breaches in the basement membrane where neurones over-migrate and exit the cortical plate.

The interaction between RG end-feet and the BM is essential to maintain the radial glial scaffold and facilitate neuronal migration even in early stages of cortical development. Consequently, retraction of RG end-feet can disrupt the position of Cajal Retzius cells (CR), a family of early postmitotic neurones of the cortical plate (Kwon et al., 2011; Meyer et al., 1999). CR cells are a heterogeneous population of neurones that migrate to the cortex from a number of locations including the septum, cortical hem and ventral pallium (Bielle et al., 2005; Takiguchi-Hayashi et al., 2004; Yoshida et al., 2006; Zhao et al., 2006). They arrange in the marginal zone (MZ) of the developing cortex, in a dispersed monolayer that lies close to the pial end-feet of radial glia (Meyer, 2010). A key function attributed to CR cells is to direct the laminar position of migrating glutamatergic neurones through the actions of a secreted glycoprotein, reelin (Alcantara et al., 1998; Chai et al., 2009; Luque et al., 2003; Meyer, 2010; Meyer et al., 1999). Loss of reelin expression or displacement of CR cells can have a major impact on the correct lamination of the neocortex and is associated with severe developmental defects in rodents and humans, most often manifest as classical type-I lissencephaly (D'Arcangelo, 2006; Frotscher et al., 2009; Hartfuss et al., 2003; Honda et al., 2011; Katsuyama and Terashima, 2009). However, the multiple sources of CR cells as well as the secretion of reelin from additional sources have complicated the clarification of their specific roles, as recently illustrated by genetic ablation of CR cells from the cortical hem which failed to cause detectable lamination defects in mice (Yoshida et al., 2006).

RG also play the role of progenitors and are thought to produce all glutamatergic neurones of the foetal cortical plate (Gotz and Huttner, 2005). Postnatally they are the main source of glia, primarily astrocytes (Costa et al., 2009; Gritti and Bonfanti, 2007; Miller and Gauthier, 2007). Unexpectedly, a recent report suggested even broader progenitor capabilities of RG. In humans, but not in rodents, progenitor cells co-expressing the RG marker, calretinin and GABA the general marker of cortical interneurones, were observed. This would mean that in humans the progenitor capability of RG may also include the generation of interneurones as well as glutamatergic neurones and glia (Howard et al., 2008; Yu and Zecevic, 2011).

Based on the available evidence we can conclude that maintaining close communication between RG, early postmitotic neurones and the BM are key to determining the timing and organisation of cortical development.

SIGNALLING LINKS BETWEEN THE BM AND THE CYTOSKELETON OF CORTICAL CELLS

The ability of migrating neurones and supporting RG to dynamically modify their morphology is crucial for normal cortical development. It therefore comes as no surprise that changes in the organisation of the neuronal or glial cytoskeleton can disrupt the laminar organisation of the developing cerebral cortex (Friocourt et al., 2011; Heng et al., 2010). In truth, the number of signalling proteins that have been identified to have essential roles in neuronal migration has increased so fast as to somewhat remove the novelty normally associated with such findings. In most cases pyramidal neurones that cannot dynamically organise their F-actin filaments and microtubules are unable to migrate through the cortical plate and arrest beneath it. In contrast, cytoskeletal changes that are linked with exencephaly in the developing cerebral cortex are far less common and thus may be useful tools to understand the molecular pathways that enable communication between the BM, RG and migrating neurones.

Studies of GPR56, a seven transmembrane G protein coupled receptor revealed that G protein signalling from radial glia end-feet is required for the maintenance of an intact BM. Mice lacking GPR56 expression initially formed a BM, however, it subsequently broke up, causing loss of RG attachment to the pial surface, displacement of CR cells, neuronal ectopias and a bumpy appearance of the cortical surface often referred to as cobblestone lissencephaly (Figure 1C) (Li et al., 2008). A putative GPR56 ligand was identified in the marginal zone and extracellular matrix which overlies the cerebral cortex. Therefore, signalling from GPR56 is likely triggered when RG end-feet come into contact with the cortical periphery. Interestingly, mutations in the *GPR56* gene have been linked with bilateral frontoparietal polymicrogyria (BFPP), a recessive genetic disorder in humans (Piao et al., 2005). Since the histopathology of BFPP is not fully clear, the biological consequences of GPR65 loss in mice have provided considerable insight into the pathological mechanisms of this disease.

Further confirmation that G protein signalling is important for BM integrity came from studies of the α subunits of heterotrimeric G proteins 12 and 13 (Gα12 and Gα13). Deletion of *Gα12* and *Gα13* was recently associated with neuronal over-migration in the cerebral cortex, suggesting their possible involvement in cobblestone lissencephaly (Moers et al., 2008). Loss of Gα12 and Gα13 in neuronal progenitors was achieved using nestin-driven Cre recombination causing BM disruptions, detachment of RG from the BM and neuronal ectopias. Initially, these phenotypes seemed to confirm that BM breakage causes cobblestone lissencephaly, as widely reported. However, detailed analyses revealed the presence of ectopic neurones that had over-migrated into the MZ as early as embryonic day 15.5 (E15.5), prior to the appearance of BM defects. Fragmentation of the basal lamina was first observed after E16.5 and increased in severity with age. Furthermore, no displacement of CR cells was apparent, the correct number of which remained evident even on the surface of the neuronal ectopias. These observations suggested that cobblestone lissencephaly may also be caused by the inability of migrating neurones to receive and/or process STOP signals. The authors identified lisophospholipids LPA and SP1 as likely extracellular stop signals that normally prevent neuronal over-migration in the cortex by activating Gα12 and Gα13. Lisophospholipids commonly affect the function of small GTPases of the Rho family, which can be activated through heterotrimeric G protein signalling and are widely known as key regulators of the cellular cytoskeleton, particularly neurones (Auer et al., 2011; Govek et al., 2011; Mack et al., 2011; Moolenaar et al., 2004). A novel insight was thus made into the molecular mechanism utilised by migrating glutamatergic neurones when approaching their final destination in the cerebral cortex.

Additional evidence supporting the role of small GTPases was gained following genetic mutation of C3G, an activating, guanine nucleotide exchange factor for the Ras family of GTPases, predominantly Rap1 (Ohba et al., 2001). Mutant mice harbouring a hypomorphic allele of *C3G* (*C3G gt/gt*) that expresses less than 5% of normal protein levels, exhibited severe disruptions in cortical development primarily involving defective organisation of the BM and the inability of neurones to migrate correctly (Voss et al., 2008). Many neurones failed to migrate into the cortical plate and arrested in the IZ with multipolar morphologies. The preplate did not split normally so the marginal zone and subplate were not formed, causing a similar phenotype to that seen in reeler mice. Interestingly, activation of C3G can be induced by exposure to reelin, subsequently causing GTP loading of Rap1 (Ballif et al., 2004; Voss et al., 2008). Reduced levels of activated Rap1 were observed in C3G mutant cells, further confirming the signalling links between these proteins required for cortical development (Voss et al., 2008). Interestingly, C3G also has reelin independent functions since *C3G gt/gt* mice exhibited additional disruptions in CR cell distribution, discontinuous laminin deposition, disrupted BM integrity and loss of RG end-feet anchoring, normally associated with the loss of integrin, rather than reelin signalling. Since β1-integrins can activate C3G, it is likely that in the developing cortex, C3G represents a converging point for the reelin and integrin signalling pathways (Arai et al., 2001; Arai et al., 1999; Uemura and Griffin, 1999).

Cytoskeletal proteins such as the tubulin superfamily are considered so important for a multitude of cellular processes that it seemed unlikely that specific mutations of individual tubulin proteins would be associated with lissencephalies in humans. However, two recent breakthroughs have demonstrated just that (Jaglin et al., 2009). A hyperactive N-ethyl-N-nitrosourea (ENU)-induced genetic screen for mutations that cause developmental defects in cortical formation revealed that mutations in the guanosine triphosphate (GTP) binding pocket of α-1 tubulin (*Tuba1*) gene cause arrested migration of neurones in mouse models, reminiscent of type-I lissencephaly in humans (Keays et al., 2007). *De novo* mutations of the human orthologue (*TUBA3*) were detected in eight patients with cortical malformations of varying degrees of severity. The most serious case showed classic type-I lissencephaly with a thick disorganized cortex and suffered from severe epilepsy, mental retardation, and motor deficits (Figure 1C) (Keays et al., 2007). Indications that genetic alterations in the tubulin superfamily genes may account for some cases of type-II cobblestone lissencephaly came from studies of the *TUBB2B* gene that encodes for β-tubulin (Jaglin and Chelly, 2009). *De novo* mutations were discovered in one embryo and four unrelated patients exhibiting bilateral, asymmetrical, and anteriorly predominant polymicrogyria, dysmorphic basal ganglia and corpus callosum agenesis or dysgenesis. Further studies of the embryo revealed the presence of disorganized RG fibres extending beyond the BM, which was used as evidence linking *TUBB2B* loss with type-II lissencephaly. However, subsequent *TUBB2B* mRNA knockdown studies in mice caused glutamatergic neurones to arrest in the SVZ/IZ with no evidence of ectopias, mimicking type-I rather than cobblestone lissencephaly. The lack of a cobblestone phenotype in mice however does not necessarily preclude changes in β-tubulin promoting type-II lissencephaly in humans. In any case it is now certainly clear that mutations in *TUBA3* and *TUBB2B* can cause cortical lamination defects in humans.

Organisation/Function of Basement Membranes

Basement membranes (BM) are thin sheets of extracellular matrix proteins including collagen (types IV and XVIII), laminin, nidogen and heparin sulphate proteoglycans, such as perlecan and agrin (Erickson and Couchman, 2000; Timpl and Brown, 1996). A number of different isoforms of these key components exist which serve to ensure that the basement membrane is structurally and functionally adapted to its role in different tissues. In the cortex a number of BMs exist, for example major blood vessels are surrounded by both endothelial and astroglial BMs (the latter being a consequence of their being bordered by astrocyte end-feet) (Hallmann et al., 2005). The epithelium of the meninges contributes to the astroglial basement membrane and the two are collectively referred to as the parenchymal basement membrane, as they delineate the border to the brain parenchyma (Hallmann et al., 2005). The endothelial BM contains both laminin α4 and α5 chains, whilst the parenchymal basement membrane contains laminin α1 and α2 chains (Sixt et al., 2001). All BMs function as a physical barrier and the pial basement membrane is no exception in its role as a signalling module to STOP migrating neurones from exiting the brain.

In the foetal telencephalon, the integrity and structure of the pial BM is tightly linked with that of the meningeal cells that overlie it and are at least in part responsible for its formation (Siegenthaler and Pleasure, 2011). Early supporting evidence came from chemical ablation of the meningeal cells in hamsters, causing irreparable focal gaps in the BM of treated cerebella (Sievers et al., 1994). The link between meninges, the BM and cortical development have since been elegantly demonstrated using molecular and genetic approaches. Thus, genetic ablation of the forkhead transcription factor *Foxc1* in mice caused extensive loss of meningeal fibroblasts from their normal position, above the cerebral cortex. The ensuing consequences included loss of BM, poor attachment of RG and cortical dysplasia reminiscent of cobblestone lissencephaly (Hecht et al., 2010; Zarbalis et al., 2007). Further evidence was seen following genetic ablation of *Zic1* and *Zic3* expression, two members of the Zic family of zinc-finger transcription factors that are required for the proliferation of meningeal fibroblasts. The consequences of the double genetic mutants in mice included a thinner and disorganised BM devoid of several protein components, particularly laminin γ1, nidogen-1, Foxc1, and perlecan, causing a phenotype reminiscent of cobblestone type-II lissencephaly (Inoue et al., 2008). Meninges can also affect cortical development irrespective of their link with the BM. They are required as a physical substrate for migrating CR cells from the cortical hem to the neocortex. Furthermore, they express the chemokine CXCL12 which attracts and promotes the migration of CR cells, maintaining them at the cortical surface (Borrell and Marin, 2006). Together these observations indicate that communication between the meninges, BM and RG are pivotal in determining the timing, identity and migration routes of cortical neurones in the developing telencephalon.

In addition to controlling proper neuronal positioning, the composition of the BM and the ability of RG to make contacts with it can affect neuronal fate. In a way this should not come as a surprise, since RG are the main progenitors of cortical glutamatergic neurones and the contacts they make with the BM are likely to influence their course of differentiation (Costa et al., 2009; Gotz and Huttner, 2005; Gritti and Bonfanti, 2007; Miller and Gauthier, 2007). Confirmation of this came from a recent study where the authors that originally identified the link between perlecan loss-of-expression and neuronal ectopias, also revealed its effects on neuronal fate in the developing cortex (Costell et al., 1999). Costell and colleagues made use of the fact that 45% of perlecan-null embryos have milder ectopias where the neurones remain in the MZ causing cortical layer distortions, but no exencephaly. A reduced number and proliferation rate of neuronal progenitors was found to result in significant microcephaly of the perlecan-null embryos and was suggested to be due to the reduced availability of sonic hedgehog (SHH), particularly in the ventral forebrain (Giros et al., 2007).

In another study Haubst *et. al.* examined mice with targeted deletion of nidogen binding site within the laminin γ1 chain (γ1III4) which exhibit extensive retractions of RG end-feet in the cerebral cortex (Haubst et al., 2006). Unexpectedly, the authors did not observe defects in RG fate or proliferation rates, however they noted a change in neuronal fate with the significant loss of Math-2 positive pyramidal neurones that should normally localise to the outer layers of the cerebral cortex. Misplacement of Gad65 positive interneurones was also evident (Haubst et al., 2006). Interestingly, some proliferating progenitors were observed within the cortical parenchyma, suggesting that exencephaly was not restricted to postmitotic neurones but could also affect their precursors.

Glycosylation and the Development of the Neocortex

Numerous genetic mouse models with mutations in BM proteins (laminin α5, laminin γ1 and perlecan), neuronal and RG matrix receptors (integrins β1, α6, dystroglycan) or downstream signalling molecules (integrin-linked kinase and focal adhesion kinase) (Beggs et al., 2003; Belvindrah et al., 2007; Graus-Porta et al., 2001; Kwon et al., 2011; Marchetti et al., 2010; Niewmierzycka et al., 2005) underscore the importance of maintaining contacts between the RG end-feet and the pial BM. When this interaction is perturbed there is an over-migration of cortical neurones into the subarachnoid space, causing cobblestone, type-II lissencephaly, which is associated with foetal or early postnatal mortality in humans.

In the last ten years an interesting new dimension has been added with the identification of a number of forms of congenital muscular dystrophy (CMD) linked with mutations in genes that encode for proteins implicated in the glycosylation and/or processing of α dystroglycan (Barresi and Campbell, 2006; Michele and Campbell, 2003; Muntoni et al., 2004). Mutations in *POMT1* (Beltran-Valero de Bernabe et al., 2002), *POMT2* (van Reeuwijk et al., 2005), *POMGnT1* (Yoshida et al., 2001), *LARGE* (Grewal and Hewitt, 2002; Longman et al., 2003; van Reeuwijk et al., 2007), *FKRP* (Beltran-Valero de Bernabe et al., 2004; Brockington et al., 2001a; Brockington et al., 2001b) and *Fukutin (Toda, 1999)* are typically associated with a wide clinical spectrum. These include severe forms of CMD with structural brain defects as exemplified by Walker Warburg syndrome (WWS) OMIM 236670, Muscle Eye Brain disease (MEB) OMIM 253280, Fukuyama CMD (FCMD) OMIM 253800 and congenital muscular dystrophy type 1D (MDC1D) OMIM 608840). In Caucasian populations mutations are most frequently seen in the *FKRP* gene, although only a proportion of these patients have structural brain involvement (Muntoni et al., 2008). The characteristic and diagnostic feature of all of these disorders, irrespective of the gene that is mutated, is the hypoglycosylation of α dystroglycan (ADG), hence their unifying term "dystroglycanopathies" (Clement et al., 2008; Godfrey et al., 2007; Muntoni et al., 2007; Toda et al., 2003). Interestingly a large proportion of patients with a deficiency in ADG glycosylation have no mutation in either *POMT1, POMT2, POMGnT1, FKRP* or *LARGE,* implying the involvement of other genes that with recent advances in sequencing will no doubt be identified in the near future.

In order to expediate the development of therapeutic intervention in these diseases, whether targeted at the muscle, eye or brain phenotype, it is necessary to understand the process by which alterations in dystroglycan glycosylation affects basement membrane structure. Dystroglycan was initially identified as the central component of the dystrophin associated protein complex and has since been attributed with a primary role in the deposition, organisation and turnover of basement membranes (Durbeej et al., 1995; Henry and Campbell, 1998; Henry et al., 1998; Michele et al., 2002; Yurchenco et al., 2004a;

Yurchenco et al., 2004b). A and β-dystroglycan are encoded by the *DAG1* gene; whilst both are glycosylated it is the extensive O-linked mannosylation of the central mucin domain of α-dystroglycan that is essential for the binding to LG domain containing proteins such as laminin (Ervasti and Campbell, 1993), perlecan (Peng et al., 1998) agrin (Bowe et al., 1994; Campanelli et al., 1994; Gee et al., 1994) and neurexin in the brain (Sugita et al., 2001), as well as pikachurin in the eye (Hohenester et al., 1999; Sato et al., 2008; Tisi et al., 2000). B-dystroglycan (BDG) interacts with dystrophin, utrophin, caveolin and Grb2 (Yang et al., 1995) thus linking the cytoskeleton with the extracellular matrix via α-dystroglycan. Whilst detailed structural knowledge regarding the pattern of α-dystroglycan glycosylation is lacking, POMT1/POMT2 and POMGnT1 are known to facilitate the initial steps of glycan addition(Endo, 2007). More recently the presence of a phosphorylated O-mannosyl glycan on the mucin-like domain has been identified (Yoshida-Moriguchi et al., 2010). This modification is thought to be mediated by LARGE and has been shown to be necessary for the laminin binding properties of α-dystroglycan (Yoshida-Moriguchi et al., 2010).

A number of animal models have now been used to investigate various aspects of the dystroglycanopathies. One of the first to be generated was the dystroglycan null mouse (*Dag1*-null) which demonstrated the crucial role that both α and β-dystroglycan play in early basement membrane formation. BM abnormalities were evident from E6.5 when a structurally defined Reichert's membrane was lacking (Williamson et al., 1997). Nulls for POMT1 (Willer et al., 2004), POMT2 (Hu et al., 2011), fukutin (Kurahashi et al., 2005) and fukutin related protein (Chan et al., 2010) have subsequently been generated, all of which caused embryonic lethality, confirming the essential role played by α-dystroglycan glycosylation in the formation of the early BMs. Whilst these findings all demonstrate a role for dystroglycan and its proper glycosylation during early development, it was a series of conditional dystroglycan null mice which identified dystroglycan as a central player in the dystroglycanopathy phenotype. Two studies have confirmed that simultaneous deletion of *DAG1* in neurones and glia causes a number of neuronal migration defects including, ectopic cells and laminar disorganisation (Moore et al., 2002; Satz et al., 2010). Interestingly, macrocephaly was observed at an incidence of 20-30%, suggesting that dystroglycan is required to regulate progenitor proliferation in the cerebral cortex (Moore et al., 2002; Satz et al., 2010).

More recently, expression of Cre recombinase from a nestin promoter was used to specifically remove *DAG1* from RG progenitors and thus all their neuronal and glial progeny. The consequences of this included extensive glial and neuronal heterotopias over the entire cerebral convexities; extending into leptomeninges and obliterating the subarachnoid space. The BM was disrupted as indicated by the discontinuous distribution of laminin, parts of which could be seen in the brain parenchyma, deep underneath the brain surface where they made contact with RG end-feet (Satz et al., 2010). The phenotype was highly reminiscent of cobblestone lissencephaly, demonstrating the central role of dystroglycan in some cases of type-II lissencephaly in humans. Interestingly a comparison of nestin-, GFAP- or NEX-driven Cre mediated deletion suggested that whilst neuronal dystroglycan plays a role in synaptic plasticity, glial dystroglycan is primarily involved in forebrain development, demonstrated by the appearance of exencephaly in *GFAP-Cre/DG* mice. This diversity in function may underlie the broad spectrum of brain involvement seen in dystroglycanopathy patients (Mercuri et al., 2009). Finally, the requirement for dystroglycan in cortical development primarily depends on its extracellular interactions rather than intracellular signalling, since deletion of the cytoplasmic domain of β-dystroglycan had no detrimental effects on neuronal migration or organisation of the glia limitans (Satz et al., 2010).

These observations indicate a degree of functional diversity with respect to the role of dystroglycan in neurones and glia. The latter aspect should perhaps come as no surprise given that the extent of ADG glycosylation has previously been shown to differ between tissues (Ervasti and Campbell, 1993). Indeed, FKRP knock down mice show tissue specific differences in basement membrane formation which appear to have a clear impact on the observed tissue specific phenotype ((Ackroyd et al., 2009); Ackroyd and Brown, unpublished data). Overall these observations clearly demonstrate the central role of α-dystroglycan glycosylation in maintaining RG:BM interactions.

Potential Therapeutic Targets

Historically, the general clinical diagnosis of lissencephaly was given to patients with apparently smooth brains (as emphasised by the Greek meaning of the words *lissos*-smooth and *cephaly*-brain). Extensive research has since demonstrated that lissencephalies are caused by a large number of different genetic defects, all of which manifest early during embryonic development. Consequential issues of diagnosis linked with appropriate, specific treatments pose a formidable task not least because therapy might be required for extended periods during the critical stages of brain development. A second major hurdle to overcome is drug delivery, as the targeting of the correct cell types would be essential if normal lamination was to be restored. Despite these obvious challenges various strategies are being put forward, including the possibility of effectively reinitiating developmental programs postnatally with the aim to restore some aspects of cortical lamination, as shown recently in a rat model of subcortical band heterotopia (Manent et al., 2009). Other approaches have included the augmentation of LIS1 protein levels via calpain inhibitors which partially rescued the *in vivo* phenotype of a mouse model of lissencephaly (Yamada et al., 2009). The restoration of ADG glycosylation via the up-regulation of LARGE has also been proposed. From a theoretical point of view this approach would seem to be particularly advantageous as LARGE is able to bypass defects of other glycosyltransferases to hyperglycosylate ADG in dystroglycan-opathies (Barresi and Campbell, 2006). Mice overexpressing LARGE show no overt muscle pathology as a result of ADG hyperglycosylation. However, the situation in the brain is less clear, as expression levels were lower in this tissue despite LARGE being controlled by a ubiquitously expressed promoter (Brockington et al., 2010). Together these developments have demonstrated a way forward, where appropriate animal models can be used to better understand the molecular and cellular mechanisms of brain disorders, as well as a means to test novel therapeutic strategies.

Whilst the brain may remain a challenge for some time, considerable advances are now being made with regard to the eye. The eye represents an easier target since it can be approached as a separate compartment which facilitates the targeted delivery of potential therapeutic agents. Crucially, eye abnormalities often do not occur in isolation, but manifest together with other CNS defects, particularly in the brain. For instance, ocular abnormalities such as retinal dysplasia, ERG abnormalities, and retinal detachments are characteristic of patients with severe forms of dystroglycanopathy, particularly those suffering from Muscle

Eye Brain Disease. Specifically, loss of α-dystroglycan:pikachurin interactions at the photoreceptor ribbon synapse are thought to contribute to synaptic defects. Pikachurin binding is lost in POMGnT1 null mice but has recently been shown to be restored by the increased expression of LARGE (Hu et al., 2011), suggesting a possible means of therapy for the eye phenotype in this group of diseases. Such approaches are however, heavily dependent on efficient delivery. In this regard it is interesting to note that the delivery of AAV-mediated RPE65 expression has produced some encouraging results in cases of Leber congenital amaurosis (LCA). Stem cells have also recently been shown to be incorporated into the retina following laser injury using either intravitreal or intravenous routes of delivery (Singh et al., 2011). These examples demonstrate the advance currently made towards understanding causes and devising treatments for eye abnormalities; they in turn may provide insight into novel approaches for symptom alleviation and possible reversal of developmental defects of the brain.

REFERENCES

Ackroyd, M.R., Skordis, L., Kaluarachchi, M., Godwin, J., Prior, S., Fidanboylu, M., Piercy, R.J., Muntoni, F., and Brown, S.C. (2009). Reduced expression of fukutin related protein in mice results in a model for fukutin related protein associated muscular dystrophies. *Brain* 132, 439-451.

Alcantara, S., Ruiz, M., D'Arcangelo, G., Ezan, F., de Lecea, L., Curran, T., Sotelo, C., and Soriano, E. (1998). Regional and cellular patterns of reelin mRNA expression in the forebrain of the developing and adult mouse. *J Neurosci 18*, 7779-7799.

Arai, A., Nosaka, Y., Kanda, E., Yamamoto, K., Miyasaka, N., and Miura, O. (2001). Rap1 is activated by erythropoietin or interleukin-3 and is involved in regulation of beta1 integrin-mediated hematopoietic cell adhesion. *J Biol Chem* 276, 10453-10462.

Arai, A., Nosaka, Y., Kohsaka, H., Miyasaka, N., and Miura, O. (1999). CrkL activates integrin-mediated hematopoietic cell adhesion through the guanine nucleotide exchange factor C3G. *Blood* 93, 3713-3722.

Auer, M., Hausott, B., and Klimaschewski, L. (2011). Rho GTPases as regulators of morphological neuroplasticity. *Ann Anat.*

Ayala, R., Shu, T., and Tsai, L.H. (2007). Trekking across the brain: the journey of neuronal migration. *Cell* 128, 29-43.

Ballif, B.A., Arnaud, L., Arthur, W.T., Guris, D., Imamoto, A., and Cooper, J.A. (2004). Activation of a Dab1/CrkL/C3G/Rap1 pathway in Reelin-stimulated neurons. *Curr Biol* 14, 606-610.

Barresi, R., and Campbell, K.P. (2006). Dystroglycan: from biosynthesis to pathogenesis of human disease. *J Cell Sci* 119, 199-207.

Beggs, H.E., Schahin-Reed, D., Zang, K., Goebbels, S., Nave, K.A., Gorski, J., Jones, K.R., Sretavan, D., and Reichardt, L.F. (2003). FAK deficiency in cells contributing to the basal lamina results in cortical abnormalities resembling congenital muscular dystrophies. *Neuron* 40, 501-514.

Beltran-Valero de Bernabe, D., Currier, S., Steinbrecher, A., Celli, J., van Beusekom, E., van der Zwaag, B., Kayserili, H., Merlini, L., Chitayat, D., Dobyns, W.B., *et al.* (2002).

Mutations in the O-mannosyltransferase gene POMT1 give rise to the severe neuronal migration disorder Walker-Warburg syndrome. *Am J Hum Genet* 71, 1033-1043.

Beltran-Valero de Bernabe, D., Voit, T., Longman, C., Steinbrecher, A., Straub, V., Yuva, Y., Herrmann, R., Sperner, J., Korenke, C., Diesen, C., *et al.* (2004). Mutations in the FKRP gene can cause muscle-eye-brain disease and Walker-Warburg syndrome. *J Med Genet* 41, e61.

Belvindrah, R., Graus-Porta, D., Goebbels, S., Nave, K.A., and Muller, U. (2007). Beta1 integrins in radial glia but not in migrating neurons are essential for the formation of cell layers in the cerebral cortex. *J Neurosci 27*, 13854-13865.

Bielas, S., Higginbotham, H., Koizumi, H., Tanaka, T., and Gleeson, J.G. (2004). Cortical neuronal migration mutants suggest separate but intersecting pathways. *Annu Rev Cell Dev Biol* 20, 593-618.

Bielle, F., Griveau, A., Narboux-Neme, N., Vigneau, S., Sigrist, M., Arber, S., Wassef, M., and Pierani, A. (2005). Multiple origins of Cajal-Retzius cells at the borders of the developing pallium. *Nat Neurosci* 8, 1002-1012.

Borrell, V., and Marin, O. (2006). Meninges control tangential migration of hem-derived Cajal-Retzius cells via CXCL12/CXCR4 signaling. *Nat Neurosci* 9, 1284-1293.

Bowe, M.A., Deyst, K.A., Leszyk, J.D., and Fallon, J.R. (1994). Identification and purification of an agrin receptor from Torpedo postsynaptic membranes: a heteromeric complex related to the dystroglycans. *Neuron* 12, 1173-1180.

Brockington, M., Blake, D.J., Prandini, P., Brown, S.C., Torelli, S., Benson, M.A., Ponting, C.P., Estournet, B., Romero, N.B., Mercuri, E., *et al.* (2001a). Mutations in the fukutin-related protein gene (FKRP) cause a form of congenital muscular dystrophy with secondary laminin alpha2 deficiency and abnormal glycosylation of alpha-dystroglycan. *Am J Hum Genet* 69, 1198-1209.

Brockington, M., Torelli, S., Sharp, P.S., Liu, K., Cirak, S., Brown, S.C., Wells, D.J., and Muntoni, F. (2010). Transgenic overexpression of LARGE induces alpha-dystroglycan hyperglycosylation in skeletal and cardiac muscle. *PLoS One 5*, e14434.

Brockington, M., Yuva, Y., Prandini, P., Brown, S.C., Torelli, S., Benson, M.A., Herrmann, R., Anderson, L.V., Bashir, R., Burgunder, J.M., *et al.* (2001b). Mutations in the fukutin-related protein gene (FKRP) identify limb girdle muscular dystrophy 2I as a milder allelic variant of congenital muscular dystrophy MDC1C. *Hum Mol Genet* 10, 2851-2859.

Campanelli, J.T., Roberds, S.L., Campbell, K.P., and Scheller, R.H. (1994). A role for dystrophin-associated glycoproteins and utrophin in agrin-induced AChR clustering. *Cell* 77, 663-674.

Cayre, M., Canoll, P., and Goldman, J.E. (2009). Cell migration in the normal and pathological postnatal mammalian brain. *Prog Neurobiol 88*, 41-63.

Chai, X., Forster, E., Zhao, S., Bock, H.H., and Frotscher, M. (2009). Reelin acts as a stop signal for radially migrating neurons by inducing phosphorylation of n-cofilin at the leading edge. *Commun Integr Biol* 2, 375-377.

Chan, Y.M., Keramaris-Vrantsis, E., Lidov, H.G., Norton, J.H., Zinchenko, N., Gruber, H.E., Thresher, R., Blake, D.J., Ashar, J., Rosenfeld, J., *et al.* (2010). Fukutin-related protein is essential for mouse muscle, brain and eye development and mutation recapitulates the wide clinical spectrums of dystroglycanopathies. *Hum Mol Genet* 19, 3995-4006.

Clement, E.M., Godfrey, C., Tan, J., Brockington, M., Torelli, S., Feng, L., Brown, S.C., Jimenez-Mallebrera, C., Sewry, C.A., Longman, C., *et al.* (2008). Mild POMGnT1 mutations underlie a novel limb-girdle muscular dystrophy variant. *Arch Neurol* 65, 137-141.

Costa, M.R., Bucholz, O., Schroeder, T., and Gotz, M. (2009). Late origin of glia-restricted progenitors in the developing mouse cerebral cortex. *Cereb Cortex* 19 Suppl 1, i135-143.

Costell, M., Gustafsson, E., Aszodi, A., Morgelin, M., Bloch, W., Hunziker, E., Addicks, K., Timpl, R., and Fassler, R. (1999). Perlecan maintains the integrity of cartilage and some basement membranes. *J Cell Biol* 147, 1109-1122.

D'Arcangelo, G. (2006). Reelin mouse mutants as models of cortical development disorders. *Epilepsy Behav* 8, 81-90.

Del Bene, F. (2011). Interkinetic nuclear migration: cell cycle on the move. *EMBO J* 30, 1676-1677.

Durbeej, M., Larsson, E., Ibraghimov-Beskrovnaya, O., Roberds, S.L., Campbell, K.P., and Ekblom, P. (1995). Non-muscle alpha-dystroglycan is involved in epithelial development. *J Cell Biol* 130, 79-91.

Endo, T. (2007). Dystroglycan glycosylation and its role in alpha-dystroglycanopathies. *Acta Myol* 26, 165-170.

Erickson, A.C., and Couchman, J.R. (2000). Still more complexity in mammalian basement membranes. *J Histochem Cytochem* 48, 1291-1306.

Ervasti, J.M., and Campbell, K.P. (1993). A role for the dystrophin-glycoprotein complex as a transmembrane linker between laminin and actin. *J Cell Biol* 122, 809-823.

Friocourt, G., Marcorelles, P., Saugier-Veber, P., Quille, M.L., Marret, S., and Laquerriere, A. (2011). Role of cytoskeletal abnormalities in the neuropathology and pathophysiology of type I lissencephaly. *Acta Neuropathol* 121, 149-170.

Frotscher, M., Chai, X., Bock, H.H., Haas, C.A., Forster, E., and Zhao, S. (2009). Role of Reelin in the development and maintenance of cortical lamination. *J Neural Transm* 116, 1451-1455.

Gee, S.H., Montanaro, F., Lindenbaum, M.H., and Carbonetto, S. (1994). Dystroglycan-alpha, a dystrophin-associated glycoprotein, is a functional agrin receptor. *Cell* 77, 675-686.

Giros, A., Morante, J., Gil-Sanz, C., Fairen, A., and Costell, M. (2007). Perlecan controls neurogenesis in the developing telencephalon. *BMC Dev Biol* 7, 29.

Godfrey, C., Clement, E., Mein, R., Brockington, M., Smith, J., Talim, B., Straub, V., Robb, S., Quinlivan, R., Feng, L., *et al.* (2007). Refining genotype phenotype correlations in muscular dystrophies with defective glycosylation of dystroglycan. *Brain* 130, 2725-2735.

Gotz, M., and Huttner, W.B. (2005). The cell biology of neurogenesis. *Nat Rev Mol Cell Biol* 6, 777-788.

Govek, E.E., Hatten, M.E., and Van Aelst, L. (2011). The role of Rho GTPase proteins in CNS neuronal migration. *Dev Neurobiol* 71, 528-553.

Graus-Porta, D., Blaess, S., Senften, M., Littlewood-Evans, A., Damsky, C., Huang, Z., Orban, P., Klein, R., Schittny, J.C., and Muller, U. (2001). Beta1-class integrins regulate the development of laminae and folia in the cerebral and cerebellar cortex. *Neuron* 31, 367-379.

Grewal, P.K., and Hewitt, J.E. (2002). Mutation of Large, which encodes a putative glycosyltransferase, in an animal model of muscular dystrophy. *Biochim Biophys Acta* 1573, 216-224.

Gritti, A., and Bonfanti, L. (2007). Neuronal-glial interactions in central nervous system neurogenesis: the neural stem cell perspective. *Neuron Glia Biol* 3, 309-323.

Hallmann, R., Horn, N., Selg, M., Wendler, O., Pausch, F., and Sorokin, L.M. (2005). Expression and function of laminins in the embryonic and mature vasculature. *Physiol Rev* 85, 979-1000.

Hartfuss, E., Forster, E., Bock, H.H., Hack, M.A., Leprince, P., Luque, J.M., Herz, J., Frotscher, M., and Gotz, M. (2003). Reelin signaling directly affects radial glia morphology and biochemical maturation. *Development* 130, 4597-4609.

Haubst, N., Georges-Labouesse, E., De Arcangelis, A., Mayer, U., and Gotz, M. (2006). Basement membrane attachment is dispensable for radial glial cell fate and for proliferation, but affects positioning of neuronal subtypes. *Development* 133, 3245-3254.

Hecht, J.H., Siegenthaler, J.A., Patterson, K.P., and Pleasure, S.J. (2010). Primary cellular meningeal defects cause neocortical dysplasia and dyslamination. *Ann Neurol* 68, 454-464.

Heng, J.I., Chariot, A., and Nguyen, L. (2010). Molecular layers underlying cytoskeletal remodelling during cortical development. *Trends Neurosci* 33, 38-47.

Henry, M.D., and Campbell, K.P. (1998). A role for dystroglycan in basement membrane assembly. *Cell* 95, 859-870.

Henry, M.D., Williamson, R.A., and Campbell, K.P. (1998). Analysis of the role of dystroglycan in early postimplantation mouse development. *Ann N Y Acad Sci* 857, 256-259.

Hohenester, E., Tisi, D., Talts, J.F., and Timpl, R. (1999). The crystal structure of a laminin G-like module reveals the molecular basis of alpha-dystroglycan binding to laminins, perlecan, and agrin. *Mol Cell* 4, 783-792.

Honda, T., Kobayashi, K., Mikoshiba, K., and Nakajima, K. (2011). Regulation of Cortical Neuron Migration by the Reelin Signaling Pathway. *Neurochem Res*. in press

Howard, B.M., Zhicheng, M., Filipovic, R., Moore, A.R., Antic, S.D., and Zecevic, N. (2008). Radial glia cells in the developing human brain. *Neuroscientist* 14, 459-473.

Hu, H., Li, J., Gagen, C.S., Gray, N.W., Zhang, Z., Qi, Y., and Zhang, P. (2011). Conditional knockout of protein O-mannosyltransferase 2 reveals tissue-specific roles of O-mannosyl glycosylation in brain development. *J Comp Neurol* 519, 1320-1337.

Inoue, T., Ogawa, M., Mikoshiba, K., and Aruga, J. (2008). Zic deficiency in the cortical marginal zone and meninges results in cortical lamination defects resembling those in type II lissencephaly. *J Neurosci* 28, 4712-4725.

Jaglin, X.H., and Chelly, J. (2009). Tubulin-related cortical dysgeneses: microtubule dysfunction underlying neuronal migration defects. *Trends Genet 25*, 555-566.

Jaglin, X.H., Poirier, K., Saillour, Y., Buhler, E., Tian, G., Bahi-Buisson, N., Fallet-Bianco, C., Phan-Dinh-Tuy, F., Kong, X.P., Bomont, P., *et al.* (2009). Mutations in the beta-tubulin gene TUBB2B result in asymmetrical polymicrogyria. *Nat Genet* 41, 746-752.

Katsuyama, Y., and Terashima, T. (2009). Developmental anatomy of reeler mutant mouse. *Dev Growth Differ* 51, 271-286.

Keays, D.A., Tian, G., Poirier, K., Huang, G.J., Siebold, C., Cleak, J., Oliver, P.L., Fray, M., Harvey, R.J., Molnar, Z., *et al.* (2007). Mutations in alpha-tubulin cause abnormal neuronal migration in mice and lissencephaly in humans. *Cell* 128, 45-57.

Kurahashi, H., Taniguchi, M., Meno, C., Taniguchi, Y., Takeda, S., Horie, M., Otani, H., and Toda, T. (2005). Basement membrane fragility underlies embryonic lethality in fukutin-null mice. *Neurobiol Dis* 19, 208-217.

Kwon, H.J., Ma, S., and Huang, Z. (2011). Radial glia regulate Cajal-Retzius cell positioning in the early embryonic cerebral cortex. *Dev Biol* 351, 25-34.

Li, S., Jin, Z., Koirala, S., Bu, L., Xu, L., Hynes, R.O., Walsh, C.A., Corfas, G., and Piao, X. (2008). GPR56 regulates pial basement membrane integrity and cortical lamination. *J Neurosci* 28, 5817-5826.

Longman, C., Brockington, M., Torelli, S., Jimenez-Mallebrera, C., Kennedy, C., Khalil, N., Feng, L., Saran, R.K., Voit, T., Merlini, L., *et al.* (2003). Mutations in the human LARGE gene cause MDC1D, a novel form of congenital muscular dystrophy with severe mental retardation and abnormal glycosylation of alpha-dystroglycan. *Hum Mol Genet* 12, 2853-2861.

Luque, J.M., Morante-Oria, J., and Fairen, A. (2003). Localization of ApoER2, VLDLR and Dab1 in radial glia: groundwork for a new model of reelin action during cortical development. *Brain Res Dev Brain Res* 140, 195-203.

Mack, N.A., Whalley, H.J., Castillo-Lluva, S., and Malliri, A. (2011). The diverse roles of Rac signaling in tumorigenesis. *Cell Cycle* 10.

Manent, J.B., Wang, Y., Chang, Y., Paramasivam, M., and LoTurco, J.J. (2009). Dcx reexpression reduces subcortical band heterotopia and seizure threshold in an animal model of neuronal migration disorder. *Nat Med* 15, 84-90.

Marchetti, G., Escuin, S., van der Flier, A., De Arcangelis, A., Hynes, R.O., and Georges-Labouesse, E. (2010). Integrin alpha5beta1 is necessary for regulation of radial migration of cortical neurons during mouse brain development. *Eur J Neurosci* 31, 399-409.

Marin, O., and Rubenstein, J.L. (2003). Cell migration in the forebrain. *Annu Rev Neurosci* 26, 441-483.

Mercuri, E., Messina, S., Bruno, C., Mora, M., Pegoraro, E., Comi, G.P., D'Amico, A., Aiello, C., Biancheri, R., Berardinelli, A., *et al.* (2009). Congenital muscular dystrophies with defective glycosylation of dystroglycan: a population study. *Neurology* 72, 1802-1809.

Merot, Y., Retaux, S., and Heng, J.I. (2009). Molecular mechanisms of projection neuron production and maturation in the developing cerebral cortex. *Semin Cell Dev Biol 20*, 726-734.

Metin, C., Baudoin, J.P., Rakic, S., and Parnavelas, J.G. (2006). Cell and molecular mechanisms involved in the migration of cortical interneurons. *Eur J Neurosci 23*, 894-900.

Metin, C., Vallee, R.B., Rakic, P., and Bhide, P.G. (2008). Modes and mishaps of neuronal migration in the mammalian brain. *J Neurosci* 28, 11746-11752.

Meyer, G. (2010). Building a human cortex: the evolutionary differentiation of Cajal-Retzius cells and the cortical hem. *J Anat* 217, 334-343.

Meyer, G., Goffinet, A.M., and Fairen, A. (1999). What is a Cajal-Retzius cell? A reassessment of a classical cell type based on recent observations in the developing neocortex. *Cereb Cortex* 9, 765-775.

Michele, D.E., Barresi, R., Kanagawa, M., Saito, F., Cohn, R.D., Satz, J.S., Dollar, J., Nishino, I., Kelley, R.I., Somer, H., *et al.* (2002). Post-translational disruption of dystroglycan-ligand interactions in congenital muscular dystrophies. *Nature* 418, 417-422.

Michele, D.E., and Campbell, K.P. (2003). Dystrophin-glycoprotein complex: post-translational processing and dystroglycan function. *J Biol Chem* 278, 15457-15460.

Miller, F.D., and Gauthier, A.S. (2007). Timing is everything: making neurons versus glia in the developing cortex. *Neuron* 54, 357-369.

Mochida, G.H., and Walsh, C.A. (2004). Genetic basis of developmental malformations of the cerebral cortex. *Arch Neurol* 61, 637-640.

Moers, A., Nurnberg, A., Goebbels, S., Wettschureck, N., and Offermanns, S. (2008). Galpha12/Galpha13 deficiency causes localized overmigration of neurons in the developing cerebral and cerebellar cortices. *Mol Cell Biol* 28, 1480-1488.

Moolenaar, W.H., van Meeteren, L.A., and Giepmans, B.N. (2004). The ins and outs of lysophosphatidic acid signaling. *Bioessays* 26, 870-881.

Moore, S.A., Saito, F., Chen, J., Michele, D.E., Henry, M.D., Messing, A., Cohn, R.D., Ross-Barta, S.E., Westra, S., Williamson, R.A., *et al.* (2002). Deletion of brain dystroglycan recapitulates aspects of congenital muscular dystrophy. *Nature* 418, 422-425.

Muntoni, F., Brockington, M., and Brown, S.C. (2004). Glycosylation eases muscular dystrophy. *Nat Med* 10, 676-677.

Muntoni, F., Brockington, M., Godfrey, C., Ackroyd, M., Robb, S., Manzur, A., Kinali, M., Mercuri, E., Kaluarachchi, M., Feng, L., *et al.* (2007). Muscular dystrophies due to defective glycosylation of dystroglycan. *Acta Myol* 26, 129-135.

Muntoni, F., Torelli, S., and Brockington, M. (2008). Muscular dystrophies due to glycosylation defects. *Neurotherapeutics* 5, 627-632.

Nakajima, K. (2007). Control of tangential/non-radial migration of neurons in the developing cerebral cortex. *Neurochem Int* 51, 121-131.

Niewmierzycka, A., Mills, J., St-Arnaud, R., Dedhar, S., and Reichardt, L.F. (2005). Integrin-linked kinase deletion from mouse cortex results in cortical lamination defects resembling cobblestone lissencephaly. *J Neurosci* 25, 7022-7031.

Ohba, Y., Ikuta, K., Ogura, A., Matsuda, J., Mochizuki, N., Nagashima, K., Kurokawa, K., Mayer, B.J., Maki, K., Miyazaki, J., *et al.* (2001). Requirement for C3G-dependent Rap1 activation for cell adhesion and embryogenesis. *EMBO J* 20, 3333-3341.

Peng, H.B., Ali, A.A., Daggett, D.F., Rauvala, H., Hassell, J.R., and Smalheiser, N.R. (1998). The relationship between perlecan and dystroglycan and its implication in the formation of the neuromuscular junction. *Cell Adhes Commun* 5, 475-489.

Piao, X., Chang, B.S., Bodell, A., Woods, K., Benzeev, B., Topcu, M., Guerrini, R., Goldberg-Stern, H., Sztriha, L., Dobyns, W.B., *et al.* (2005). Genotype-phenotype analysis of human frontoparietal polymicrogyria syndromes. *Ann Neurol* 58, 680-687.

Rakic, P. (2003). Developmental and evolutionary adaptations of cortical radial glia. *Cereb Cortex* 13, 541-549.

Rakic, P. (2006). A century of progress in corticoneurogenesis: from silver impregnation to genetic engineering. *Cereb Cortex* 16 Suppl *1*, i3-17.

Sato, S., Omori, Y., Katoh, K., Kondo, M., Kanagawa, M., Miyata, K., Funabiki, K., Koyasu, T., Kajimura, N., Miyoshi, T., *et al.* (2008). Pikachurin, a dystroglycan ligand, is essential for photoreceptor ribbon synapse formation. *Nat Neurosci* 11, 923-931.

Satz, J.S., Ostendorf, A.P., Hou, S., Turner, A., Kusano, H., Lee, J.C., Turk, R., Nguyen, H., Ross-Barta, S.E., Westra, S., *et al.* (2010). Distinct functions of glial and neuronal dystroglycan in the developing and adult mouse brain. *J Neurosci* 30, 14560-14572.

Siegenthaler, J.A., and Pleasure, S.J. (2011). We have got you 'covered': how the meninges control brain development. *Curr Opin Genet Dev*. in press.

Sievers, J., Pehlemann, F.W., Gude, S., and Berry, M. (1994). Meningeal cells organize the superficial glia limitans of the cerebellum and produce components of both the interstitial matrix and the basement membrane. *J Neurocytol* 23, 135-149.

Singh, T., Prabhakar, S., Gupta, A., and Anand, A. (2011). Recruitment of Stem Cells Into the Injured Retina After Laser Injury. *Stem Cells Dev*. in press.

Sixt, M., Engelhardt, B., Pausch, F., Hallmann, R., Wendler, O., and Sorokin, L.M. (2001). Endothelial cell laminin isoforms, laminins 8 and 10, play decisive roles in T cell recruitment across the blood-brain barrier in experimental autoimmune encephalomyelitis. *J Cell Biol* 153, 933-946.

Sugita, S., Saito, F., Tang, J., Satz, J., Campbell, K., and Sudhof, T.C. (2001). A stoichiometric complex of neurexins and dystroglycan in brain. *J Cell Biol* 154, 435-445.

Takiguchi-Hayashi, K., Sekiguchi, M., Ashigaki, S., Takamatsu, M., Hasegawa, H., Suzuki-Migishima, R., Yokoyama, M., Nakanishi, S., and Tanabe, Y. (2004). Generation of reelin-positive marginal zone cells from the caudomedial wall of telencephalic vesicles. *J Neurosci* 24, 2286-2295.

Tanaka, D.H., Maekawa, K., Yanagawa, Y., Obata, K., and Murakami, F. (2006). Multidirectional and multizonal tangential migration of GABAergic interneurons in the developing cerebral cortex. *Development* 133, 2167-2176.

Timpl, R., and Brown, J.C. (1996). Supramolecular assembly of basement membranes. Bioessays *18*, 123-132.

Tisi, D., Talts, J.F., Timpl, R., and Hohenester, E. (2000). Structure of the C-terminal laminin G-like domain pair of the laminin alpha2 chain harbouring binding sites for alpha-dystroglycan and heparin. *EMBO J* 19, 1432-1440.

Toda, T. (1999). [Fukutin, a novel protein product responsible for Fukuyama-type congenital muscular dystrophy]. *Seikagaku* 71, 55-61.

Toda, T., Kobayashi, K., Takeda, S., Sasaki, J., Kurahashi, H., Kano, H., Tachikawa, M., Wang, F., Nagai, Y., Taniguchi, K., *et al.* (2003). Fukuyama-type congenital muscular dystrophy (FCMD) and alpha-dystroglycanopathy. *Congenit Anom* (Kyoto) 43, 97-104.

Uemura, N., and Griffin, J.D. (1999). The adapter protein Crkl links Cbl to C3G after integrin ligation and enhances cell migration. *J Biol Chem* 274, 37525-37532.

Valcanis, H., and Tan, S.S. (2003). Layer specification of transplanted interneurons in developing mouse neocortex. *J Neurosci* 23, 5113-5122.

Valiente, M., and Marin, O. (2010). Neuronal migration mechanisms in development and disease. *Curr Opin Neurobiol* 20, 68-78.

van Reeuwijk, J., Grewal, P.K., Salih, M.A., Beltran-Valero de Bernabe, D., McLaughlan, J.M., Michielse, C.B., Herrmann, R., Hewitt, J.E., Steinbrecher, A., Seidahmed, M.Z., *et al.* (2007). Intragenic deletion in the LARGE gene causes Walker-Warburg syndrome. *Hum Genet* 121, 685-690.

van Reeuwijk, J., Janssen, M., van den Elzen, C., Beltran-Valero de Bernabe, D., Sabatelli, P., Merlini, L., Boon, M., Scheffer, H., Brockington, M., Muntoni, F., *et al.* (2005).

POMT2 mutations cause alpha-dystroglycan hypoglycosylation and Walker-Warburg syndrome. *J Med Genet* 42, 907-912.

Voss, A.K., Britto, J.M., Dixon, M.P., Sheikh, B.N., Collin, C., Tan, S.S., and Thomas, T. (2008). C3G regulates cortical neuron migration, preplate splitting and radial glial cell attachment. *Development* 135, 2139-2149.

Willer, T., Prados, B., Falcon-Perez, J.M., Renner-Muller, I., Przemeck, G.K., Lommel, M., Coloma, A., Valero, M.C., de Angelis, M.H., Tanner, W., *et al.* (2004). Targeted disruption of the Walker-Warburg syndrome gene Pomt1 in mouse results in embryonic lethality. *Proc Natl Acad Sci U S A* 101, 14126-14131.

Williamson, R.A., Henry, M.D., Daniels, K.J., Hrstka, R.F., Lee, J.C., Sunada, Y., Ibraghimov-Beskrovnaya, O., and Campbell, K.P. (1997). Dystroglycan is essential for early embryonic development: disruption of Reichert's membrane in Dag1-null mice. *Hum Mol Genet* 6, 831-841.

Yamada, M., Yoshida, Y., Mori, D., Takitoh, T., Kengaku, M., Umeshima, H., Takao, K., Miyakawa, T., Sato, M., Sorimachi, H., *et al.* (2009). Inhibition of calpain increases LIS1 expression and partially rescues in vivo phenotypes in a mouse model of lissencephaly. *Nat Med* 15, 1202-1207.

Yang, B., Jung, D., Motto, D., Meyer, J., Koretzky, G., and Campbell, K.P. (1995). SH3 domain-mediated interaction of dystroglycan and Grb2. *J Biol Chem* 270, 11711-11714.

Yokota, Y., Gashghaei, H.T., Han, C., Watson, H., Campbell, K.J., and Anton, E.S. (2007). Radial glial dependent and independent dynamics of interneuronal migration in the developing cerebral cortex. *PLoS One 2*, e794.

Yoshida-Moriguchi, T., Yu, L., Stalnaker, S.H., Davis, S., Kunz, S., Madson, M., Oldstone, M.B., Schachter, H., Wells, L., and Campbell, K.P. (2010). O-mannosyl phosphorylation of alpha-dystroglycan is required for laminin binding. *Science* 327, 88-92.

Yoshida, A., Kobayashi, K., Manya, H., Taniguchi, K., Kano, H., Mizuno, M., Inazu, T., Mitsuhashi, H., Takahashi, S., Takeuchi, M., *et al.* (2001). Muscular dystrophy and neuronal migration disorder caused by mutations in a glycosyltransferase, POMGnT1. *Dev Cell* 1, 717-724.

Yoshida, M., Assimacopoulos, S., Jones, K.R., and Grove, E.A. (2006). Massive loss of Cajal-Retzius cells does not disrupt neocortical layer order. Development *133*, 537-545.

Yu, X., and Zecevic, N. (2011). Dorsal radial glial cells have the potential to generate cortical interneurons in human but not in mouse brain. *J Neurosci* 31, 2413-2420.

Yurchenco, P.D., Amenta, P.S., and Patton, B.L. (2004a). Basement membrane assembly, stability and activities observed through a developmental lens. *Matrix Biol 22*, 521-538.

Yurchenco, P.D., Cheng, Y.S., Campbell, K., and Li, S. (2004b). Loss of basement membrane, receptor and cytoskeletal lattices in a laminin-deficient muscular dystrophy. *J Cell Sci* 117, 735-742.

Zarbalis, K., Siegenthaler, J.A., Choe, Y., May, S.R., Peterson, A.S., and Pleasure, S.J. (2007). Cortical dysplasia and skull defects in mice with a Foxc1 allele reveal the role of meningeal differentiation in regulating cortical development. *Proc Natl Acad Sci U S A 104*, 14002-14007.

Zhao, C., Guan, W., and Pleasure, S.J. (2006). A transgenic marker mouse line labels Cajal-Retzius cells from the cortical hem and thalamocortical axons. *Brain Res* 1077, 48-53.

Index

A

B

C

D

H

I

K

L

M

N

O

S

W

X

Y

Z